Defense Mechanisms

FROM VIRUS TO MAN

Defense Mechanisms
FROM VIRUS TO MAN

by HAL HELLMAN

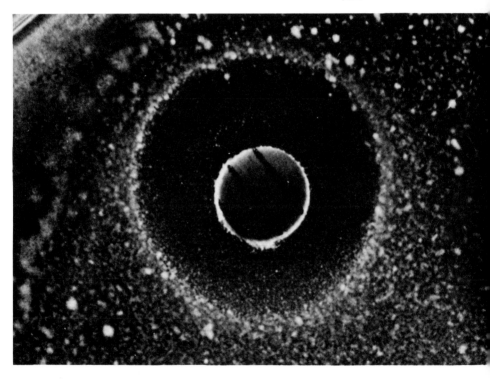

Illustrated with photographs and diagrams

HOLT, RINEHART AND WINSTON
New York   Chicago   San Francisco

PICTURE CREDITS

Allen/National Audubon Society, p. 122; Ambler/National Audubon Society, pp. 28, 64, 110, 116; Courtesy of The American Museum of Natural History, pp. 26, 73, 102; Beck/National Audubon Society, p. 43; Bennet/Monkmeyer, p. 48; Marc & Evelyn Bernheim/Rapho Guillumette, front jacket and pp. 69, 83; Chace/National Audubon Society, p. 86; Coleman & Hayward, London, p. 93; Jerry Cooke, p. 96; Walter Dawn, back jacket and pp. 55, 24; Irven DeVore, p. 115; H. E. Edgerton/National Audubon Society, p. 37; Focht-Rue/ Monkmeyer, p. 5; J. H. Gerard/Monkmeyer, p. 13; J. H. Gerard/ National Audubon Society, pp. 4, 106; Gregor/Monkmeyer, p. 81; Harrison/Monkmeyer, p. 98; Kenyon National Audubon Society, p. 108; The New Yorker, p. 135; Noailles/Rapho Guillumette, p. 88; Petersen/National Audubon Society, p. 105; G. Porter/National Audubon Society, p. 14; Roche/Monkmeyer, p. 84; Rue/Monkmeyer, pp. 7, 21, 39, 46, 49, 70, 100, 131; Rue/National Audubon Society, pp. 35, 69; Smith/National Audubon Society, p. 79; H. Spencer/National Audubon Society, p. 9; Wells/National Audubon Society, p. 93; Woolridge/Monkmeyer, p. 2; R. Wright/National Audubon Society, pp. 4, 15; Ylla/Rapho Guillumette, pp. 66, 120.

For Jack Pollock
and the memory of Lillian Pollock

The author wishes to thank the following specialists, who read all or parts of the manuscript: Dr. Morton Klass, Department of Anthropology at Barnard College and Columbia University; Dr. I. Bernard Weinstein, Associate Professor of Medicine at the College of Physicians and Surgeons of Columbia University; and Dr. Fred H. Glenny, Department of Biology at Fairleigh Dickinson University. Their comments and suggestions were very helpful.

Contents

1. Challenge and Response

THE fox glides warily through the woods, his lean body taut with hunger. There! What's that? No, just a rock.

Then he spies the outline of a plump form behind a bush, about a hundred feet away. He surveys the scene, his mouth watering in anticipation. A perfect setup, just what he's been looking for.

He moves stealthily now. Closer. Closer. Attack!

A flash of fur.

The other animal, aware that trouble is brewing, has turned his back in preparation for his defense. But no need. In the last split second before his final leap to the kill, the fox has come to a grinding halt.

The second animal, as you may have guessed, is a skunk. A bright white stripe down his back has told the fox all he needs to know. He slinks away to seek his fortunes elsewhere.

What saved the skunk's life? Obviously he was saved by his scent glands, which can squirt a powerful chemical. Right?

Wrong. In general outline the skunk looks like a number of other small animals. His scent glands would be of little use if the fox attacked him and injured or killed him before he could defend himself.

The white stripe is a flag, a warning to the predator that he is playing with fire. In other words, the skunk owes his life, in this encounter anyway, to his garish coloring rather than to his scent glands. After all, he never did fire his chemical gun.

We normally think that protective coloring means blending with the environment. But here we have seen a case where *bright* coloring serves as a defense mechanism.

Therefore, a defense mechanism is something, anything, in the physical apparatus or the way of life of a living organism that enables it to cope with a threat to its existence. Defense mechanisms are necessary because all creatures are threatened in many ways.

The World of Life

The most obvious threat is that of being injured or killed by another animal, probably a larger one. Many people have the idea that all animals are constantly running around with bloodstained claws, fighting with everything in sight. This is not the case at all. First of all, most creatures do not attack one another if not hungry. Second, many of them are herbivores, or vegetarians, and thus are not interested in meat even when they are hungry.

Most important, however, is the fact that they do not come in contact with all animals around them, but only with those in direct competition with them for food or living space. Every creature inhabits what is called an *environmental niche*. This is its way of life, or more specifically, its way of "making a living." Even though a rabbit and a squirrel may inhabit the same area and may be similar in many ways, they are not in competition. The squirrel makes his home in trees and eats nuts, acorns, and seeds; the rabbit is a burrower and eats grasses and vegetables. Thus they do not inhabit the same environmental niche.

Only when the paths of two animals cross directly, when both are in competition for the same food or mate, or when one is food for the other, do confrontation and perhaps battle ensue.

Our interest is mainly in those cases where one type of animal is normally the prey of another—the zebra and the lion, the mouse and the owl, the fly and the spider. More specifically, we are interested in how the preyed upon creatures protect themselves, and we shall see that this does not always mean fighting back.

3

Defense Mechanisms

The squirrel makes his home in trees and eats nuts, acorns, and seeds.

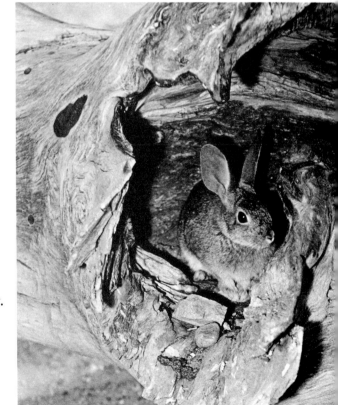

The rabbit is a burrower and eats grasses and vegetables.

Defense Mechanisms

The World Within Us

All living things are in danger of attack from inside as well as outside. The word "germ" is a general one for a number of different types of microorganisms, certain members of which cause infection and illness. Among them are bacteria, protozoa, and viruses, the last of which are the smallest and simplest organisms of all.

Most bacteria and protozoa need a moist environment to exist. The body of a large animal (particularly his mouth, throat, and intestines) provides a fine home for these microorganisms, because it supplies warmth and food as well as moisture. What more could any microorganism want? You may not like the idea, but your body —and the bodies of all multicell creatures—is filled with a multitude of these tiny organisms. Fortunately, most of them are harmless.

Indeed, some animals could not live without them. Termites, for example, have no means for digesting the wood they eat. Protozoa in their stomachs do the job for them. Goats, too, are able to digest some of the weird things they eat only because of the presence of protozoa in their stomachs.

But as we know, some of these organisms are poisonous or produce poisons within the bodies they invade. Before the days of sanitation, modern medicine, and abundant supplies of fresh food, some of these invasions occurred on a frightening scale. In the mid-fourteenth century some 25 million people perished in Europe alone under the onslaught of the Black Death. This was mainly bubonic plague (caused by the bacterium *Pasteurella pestis*), but other diseases were also involved.

6

The World Around Us

Members of the animal world may be threatened in yet another way. Their supply of water or food or oxygen may be getting low. The climate may be getting colder or warmer. In other words, the physical environment may be changing in such a way as to be a threat to the inhabitants. It is believed that the ice ages of our geological past wiped out many species of animals who were not able to cope with the changes.

Another possibility is that a species, being deficient in its means of defense against competitors, may be driven out of its normal hunting grounds and into an area that nobody else wants—into a highly mountainous region, for example. This would explain the existence of Rocky Mountain bighorn sheep in that area; it would also explain why whales and seals, both mammals, went back to the sea.

Rocky Mountain bighorn sheep.

Defense Mechanisms

Such changes in the environment may result in a physical change, sometimes called an *adaptation*, in the species. (We shall see later how this happens.) Such a change, whatever it may be, can be considered a defense mechanism. The hump of the camel permits him to store nourishment for the long treks between oases in the desert, and sweating is a defense mechanism against heat. Clearly these adaptations can be a permanent change, as in the first case, or a temporary activity, as in the second.

One of the most interesting of such defense mechanisms is found in the virus. The virus is a parasite in that it can exist only in living cells. It differs from other parasites, however, in that it has no food-processing machinery of its own.

When the virus encounters an unfavorable environment —e.g., too cold, too warm, or too dry—it simply changes to a dormant, or inactive, substance. Usually this substance takes a crystalline form, a form not normally found in the nonliving world. In this way, the virus is able to contend with the most unfavorable conditions. This ability is its defense mechanism. When it once again arrives in a favorable environment, it changes back into the active state.

Biotic Potential

All living things, including plants, have some kind of defense mechanism. There is a very simple reason for this. The basic ingredient of life is reproduction, which is necessary for the continuation of the species. But suppose every organism that was born lived and reproduced? (This potential for population growth is often called its *biotic potential*.)

Let's consider the paramecium, a protozoan. The paramecium reproduces by simply dividing in half. But this happens so often (about every thirty minutes) and so regularly that the offspring from a single one could fill the oceans, lakes, and rivers in a few months! In a very real sense, a paramecium is immortal, for it never dies a natural death. That is, every paramecium will continue to reproduce. They never grow old and die the way multicell creatures do.

But the paramecium is competing for survival with every other water-dwelling creature, which immediately limits its nutrients and living space. Also, it is food for a number of larger creatures. And finally, even if all other factors were favorable, the point would soon be reached where it would begin contaminating its living space with its own waste products. (Man is doing something similar to his own environment.)

In other words, the biotic potential is counterbalanced

Paramecia.

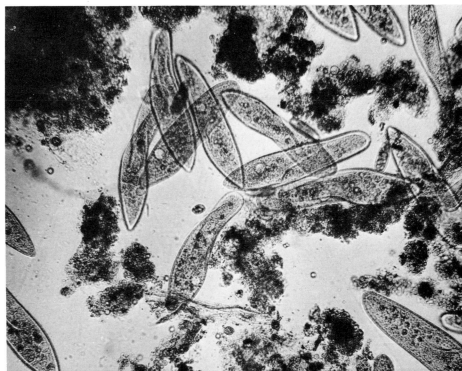

by the environmental resistance. And in general the environmental resistance is fierce. The death rate of the paramecium is so high that this protozoan would be wiped out in no time if it did not reproduce in such large numbers. Its high rate of reproduction can therefore be called a defense mechanism. And a very good one it is too.

Herbert Spencer, the great English philosopher, once said that life is "effective response." Response to what? To the challenge of the environment. And in the broad sense in which we have used the term "environment," this is challenge indeed. A defense mechanism is a response to a particular challenge.

Some of the more familiar defense mechanisms are the armor of the turtle, the camouflage of the moth, and the speed of the deer. But, as we shall soon see, some defense mechanisms are strange indeed.

2. Hide!

An excellent example of how a stronger assailant can turn predator into prey is seen in the wasp/spider relationship. The spider is normally a potent predator, catching many kinds of insects, large and small, in his web (or, more likely, *her* web; for most male spiders are harmless), injecting a paralyzing poison, and eating them.

Although predator wasps are often smaller than the spiders they go after (the bird spider of the tropics has a three and one-half inch body!), they are tough, powerful attackers who immobilize their enemies with a well-aimed sting in a nerve. This puts the spider's legs and jaw pincers out of action, after which the wasp buries it. The large, juicy body of the spider thus remains alive and fresh and provides ample nourishment for the offspring of the wasp.

The struggle between spider and wasp has been going on for millions of years. Various spiders have special means of meeting the threat. In the basic situation, the wasp swoops down like a fighter plane on the spider, who normally waits in the center of the web for an unsuspecting insect to blunder onto it and get caught. Some spiders have abandoned this vulnerable spot, taking up a position under leaves at the edge of the web. Others construct

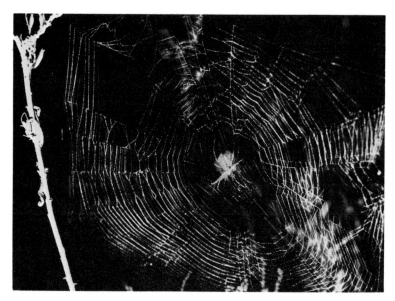

Spider vibrates himself and his web so violently they become virtually invisible.

ingenious hiding places. Some stretch alarm lines near the web. Should the wasp touch one, the spider will be given a split second of advance warning, which may be enough.

At the approach of a buzzing wasp, another kind of spider vibrates himself and his web so violently that they virtually become invisible. Try it with a piece of spring steel, or hold a knife blade down tightly at the edge of a table and flick the other end. You will see how the wasp might have difficulty deciding where to strike. Some spiders weave decoys—replicas of themselves—into their webs. Hopefully the wasp will be drawn, at least momentarily, to the decoy, giving the real creature time to escape. For this escape the spider drops quickly to earth or to leaves under which he can hide, paying out a "life line" as he goes. Without this line he could not find his way back to the web, his "home." Unfortunately for spiders, some wasps have learned to follow the fall thread and can therefore quickly trace their victims.

Things have gotten so bad that some spiders have taken to digging holes in the ground. The stronghold of the wolf spider is a well which can be as much as eighteen inches deep. When hungry he sits on his turret waiting for a meal to pass by. When one comes close enough he pursues it like a wolf and, if successful, brings it back to his lair. If

Banded garden spider.

menaced, however, he simply drops down to the safety of his stronghold. Few wasps will enter such a stronghold, for the spider will then have the advantage. Sometimes the wasp will wait at the entrance. When the spider finally clambers up, hoping that the coast is clear, the wasp seizes his forelegs, pulls him out in a high arc, and stings him.

Another kind of ground spider provides his nest with a second exit, from which he can observe the activities of the wasp at the first one. Hinged shutters or wedge-shaped plugs are sometimes used at the entrances.

One spider, *Rhytidicolus structor*, has managed to at least meet the wasp on even terms. This one builds a maze in the earth, utilizing several chambers with hinged doors. If a wasp enters the maze in search of the spider, which does occasionally happen, the spider hides in one of the chambers. At the right moment, the spider rushes out and "locks" the wasp in. When the wasp finally suffocates or dies of hunger, *Rhytidicolus structor* takes his revenge. I use such terms as "revenge" and "hope" for literary reasons.

The earth has proved a haven for many creatures, as for example, spadefoot toads. Their name is derived from the presence of a hard, bladelike projection on each rear foot.

They spend considerable time during the year several feet below ground in holes they dig for themselves. It is rather a strange sight to see a spadefoot that is facing you sink slowly out of sight as it digs backward into the ground.

In desert regions the early stages of development of these toads are extremely rapid. The tadpoles are said to hatch about thirty-six hours after the eggs are laid, and they change into small toads about twelve days later. Such rapid development permits them to exist in an environment where small pools, which they need for their early life, do not last too long. In winter they go deep underground and hibernate.

The mole, on the other hand, spends almost all his time underground. With powerful forelimbs and efficient shovel-shaped claws, he can dig a twenty-yard burrow in a single day. He does a great deal of digging in order to obtain his food. Thus, he has solved the dilemma of the small creature who wants to be concealed but who must move around to obtain his food. Obtaining air is no problem, because he usually digs his tunnels near the surface and through loosely packed earth.

In the sea, where movement of the water may bring food to the animal, other methods of concealment and protection are possible. The tiny female crab *Hapalocarcinus marsupialis* sets up housekeeping when young. She does this where two coral branches of the Great Barrier Reef, off Australia, meet. Branches grow around her, forming a cage about the size of a marble. Small holes permit fresh sea water to wash through the cage and bring food. She lives there for the rest of her life.

Another sea creature, the octopus, takes quite a different approach. I am sure that every one of you has seen at least one picture or TV program showing an octopus wrapping his powerful tentacles around a man or some other large creature. There is no question that the octopus is a frightening-looking creature, that his tentacles are powerful, and that the suckers they contain can cause trouble. But although often pictured as giants, few have arms more than four feet long, and many species have bodies no larger than a pear. These animals live in rocky places along the shore and near reefs, protecting their soft bodies by sitting in holes and crevices. They reach out with their arms to seize their victims, which are fish, crabs, lobsters, and other creatures they can overpower and eat with their horny beaks. In some cases the prey is seized by the sucker-bearing arms, paralyzed, and partially digested by a poisonous secretion. Octopuses are themselves preyed upon by large fishes, turtles, and other sea animals, plus man.

The octopus has developed several interesting means of defense, all of which involve hiding himself in one way or another. When attacked, he may squirt a brown or black fluid into the face of his enemy and then dart away to some safe area. This was probably the original smoke screen.

Further, he has color-changing cells in his skin. Indeed, of all creatures capable of changing the colors of their bodies, the octopus' color-changing cells are probably the fastest acting of all. His cells can act in seconds, whereas those of other creatures require minutes, hours, or even days. These cells are of many colors—yellow, orange, blue-green, brown. They have muscular walls that enable them to contract and expand singly or in groups, providing a great variety of possible colors.

Even his method of propulsion is interesting. It involves a kind of jet propulsion through a movable tube. The octupus takes in a quantity of water and squirts it out through the tube. By varying the direction of the outlet, he can move backward as well as forward.

In experimenting with a relative of the octopus, the squid, the English marine zoologist D. N. F. Hall was given a lesson on how color change, speed, and ink discharge can be combined to evade an attacker. Hall put a three-inch squid into a large, light-colored wooden tube and tried to catch it by hand. When his fingers were about nine inches away, the squid turned dark. Hall made a grab—and came up with a handful of ink. The squid was at the other end of the tub. He tried again, and this time was able to see what happened. After turning dark, the squid ejected ink, which simulated very roughly the shape of his own body. At the same time, he turned pale and shot away to the other end of the tub.

Hall had the advantage of seeing all this happen from above. Under natural conditions, the ink would undoubtedly have been discharged between the squid and a predator, making his escape even more likely.

Some deep-sea organisms pull the reverse trick. When pursued they eject large quantities of a luminous substance. In the darkness of the deep water the predator is temporarily blinded—just as we are when we step from darkness into sunlight—and the marine creature can make his escape.

At the opposite end of the scale we have the creatures who by their very nature are so similar in appearance (color or shape or both) to their natural environment that movement would surely give them away. Indeed, almost every preyed-upon creature seems to sense that movement is dangerous. The spotted fawn of the Virginia deer lies in trancelike immobility wherever placed by its mother after birth. It is protected because in both pattern and color it resembles very closely the background against which it is seen. This of course is an example of *camouflage*, or more precisely, protective coloration.

Another interesting example of protective coloration is that of the tree sloth, who spends his life upside down hanging from branches with his powerful hooked claws. As a matter of fact, he cannot walk or even stand upright. Thus, he has no real defenses. Running is impossible, for he can only inch along slowly, and his awkward position makes it very difficult for him to fight back. Living as he does among green leaves might normally make him quite conspicuous, for mammals never have green skins, although this color is common among insects and birds. The sloth gets around this problem by an interesting device. Green algae grow on his fur! Thus, hanging quietly, he is very difficult to distinguish from the leaves and lichen on the trees in which he makes his home.

There is another creature who lives his life upside down, the Nile catfish *Synodontis batensoda*. If we look closely at the fish, we see something else that strikes us as peculiar. His belly is darker than his back, whereas in almost all other creatures the reverse is true. The white, or light, underside is definitely the rule. This makes sense when we consider that light, under natural conditions, comes from above, which throws the underside into shadow. By having a lighter underside, the animal counteracts the rounding effect of light and shadow and thus appears flat. This effect is called *countershading* and aids birds and fish in another way. When viewed from below, they blend with the sky or the surface of the water, whereas when viewed from above, they are more closely matched with the darker land or sea bottom. If *Synodontis* swam in a normal fashion, his countershading would be all wrong.

Light and dark are used in another way too. *Disruptive coloration*, or irregular patches of contrasting colors, serve to distract the eye from the outline of the animal. The same principle has also been used to camouflage aircraft, ships and men in wartime. The ringed plover, common on the beaches of Europe, is a good example. His black color and head stripe disrupt his lines and make him blend with the background in a most unexpected way. Similarly, the tiger is very conspicuous in the zoo, but he is far less so in

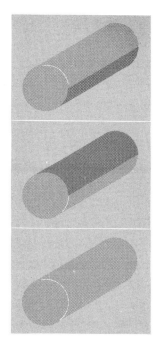

Countershaded cylinder under top lighting blending into background.

Countershaded cylinder under diffused lighting (from all sides).

Unshaded cylinder under top lighting (as from the sun).

19

the strong lights and shadows of the tall jungle grasses in which he lives.

Animals that live in varying environments are often aided by the capacity to change color. The chameleon is probably the most common example, and we mentioned the abilities of the octopus and squid earlier in the chapter. But the most amazing ability is found in certain fishes, particularly the flounder and some of its relatives, which can vary not only in color but in pattern. This shows clearly in a series of photos in which the same fish has been placed against varying backgrounds.

Another kind of color change is seen in certain animals that live in climates which change drastically with the seasons. The varying hare lives in northern regions that are blanketed by snow in the winter months. At that time his coat is snowy white. In summer, his coat changes to gray-brown, matching very closely the ground and shadows around it. White-tailed deer also change their "clothing" in winter and summer. Experiments have shown that the heat and cold have nothing to do with the changes but that the length of the days, a much more reliable indicator, is the guide. The length of time the eyes receive light determines how much of a certain hormone (chemical substance) is sent to the blood. This hormone regulates the clothing change.

Among certain species of birds, an additional bit of species-preservation magic has been added. The male members of various ducks, certain ptarmigan, and Bulwer's pheasant do not change their white winter dress until long after the females. The bright colors of the males, it would seem, deflect the attention of predators from the well-camouflaged females and chicks. The latter,

Rock ptarmigan changing from winter plumage to summer brown.

of course, are more valuable to the species because they represent the next generation.

We come now to the second main type of coloration, the revealing, as opposed to the concealing. The example of the skunk with which we opened the book is a common one. Less familiar is that of many other creatures, particularly caterpillars and butterflies.

While some butterflies are cryptically colored—that is, with concealing colors—many are brightly colored. These, with no apparent means of defense, are particularly vulnerable to many predators, ground creatures as well as birds. Or they would be, except for one thing. They taste awful. Of course, this does not protect all of them, for their major enemy, birds, must learn that butterflies with a particular color pattern are not worth bothering with. They learn this by tasting. Thus a few butterflies are sacrificed so that the rest may live.

But warning coloration must not be subtle or refined. As with the skunk, the conspicuous colors must be such as to make the process of education as rapid as possible and to make subsequent mistakes unlikely. In the domain of sound, the rattle of the rattlesnake and the hiss of other serpents serve the same function.

An interesting finding is that butterflies that are distasteful to predators are generally narrow specialists in choice of food. They tend to select plants on which other butterfly groups do not feed, particularly those rich in alkaloids, which are bitter or poisonous. Their food appears to serve a double function. It provides them with a feeding arrangement in which they have very little competition; and it may supply them with the substances that make them distasteful to predators. These butterflies ap-

parently are immune to the toxic or repellent plant substances on which they feed. They are thus able to turn the
plants' chemical defenses to their own advantage.

The monarch butterfly is one of these creatures. He is
large and brightly colored, with orange-brown wings and
black and white markings. He is the only known butterfly
that migrates from its northern haunts in Canada and the
United States to warmer regions.

The monarchs' defense is obviously a good one, for they
are found in enormous numbers. In their travels south
during the fall, it has been estimated that as many as a
billion have passed a single point. The great flocks break
up in the spring and the monarchs travel northward
again. They stop to lay their eggs on milkweed leaves
over practically all of North America from Texas to Hudson Bay.

So it seems that predators can learn to associate certain
colors and patterns with danger or distastefulness and to
leave the bearers of these colors alone. But then, what is
to prevent other animals from having these colors, thus
achieving immunity to attack by imitation of the warning
colors? The answer is that there is nothing to prevent it,
and it does indeed happen. The only requirement is that
the species that bluffs must be much less common than
the species with true warning coloration. For if the pro-
portions were the other way around, the required lesson
might never be properly learned.

Thus, there is another butterfly, the viceroy, who is a
dead ringer for the monarch. Yet he comes from a different family altogether. He does not even have the unpleasant taste of the monarch, for tame birds who have
not had unpleasant experiences with the monarch will

Viceroy (left) is protected by his close resemblance to the foul-tasting monarch (right).

feed on him. In the wild, however, the viceroy is protected by his close resemblance to the monarch, for a bird who has had experience with the taste of a monarch will leave the viceroy strictly alone. This kind of protective coloration is called *mimicry*. The species mimicked—for example, the monarch—is called the *model*, and the unprotected species is called the *mimic*.

The process not only is intraspecific (within the same species) but occurs between species as well. For example, the clear-wing moths, which are quite helpless, closely resemble bees, which are certainly not helpless and which are clearly marked.

Does it work? One observer was able to count the insects a pair of starlings fed to their young. Of more than 16,000 insects counted, fewer than 20 had warning colors!

Nor need mimicry involve only the copying of another animal or insect. Similarity to objects, such as leaves, twigs—even bird droppings!—is observable. Perhaps the most startling examples are the stick caterpillars of certain geometrid moths. These hold themselves rigid all day in amazing likeness to a twig—complete with buds—and move in search of food only when the sun goes down.

The bittern, a marsh heron, takes a rather different approach. Not only is he colored like the cattails in which he makes his home, but if a hawk or other predator appears, he stands with his body perfectly upright and stiff, his beak pointing to the sky along with the reeds around him. Further, should a breeze set the cattails swinging, the bittern sways back and forth along with them!

Some creatures even have markings that simulate their *own* bodily defenses. Many poisonous insects have what are called "bleeding points," glandular openings through which their toxic secretions ooze or are forcibly ejected. The color of the secretion is often yellow and undoubtedly forms part of the general scheme of warning coloration. However, once the insect has used up his "stock," it takes time for replenishment to take place. In the meantime he may be helpless. Consequently, various species have markings that mimic their own defensive secretions. Thus, a potential predator may be held off even though the insect is actually defenseless.

The bittern (he's really there) blends perfectly with the surrounding reeds.

Many creatures, lacking some of the more obvious means of defense, have tried to hide under cover of darkness. Indeed, as night begins to fall, a whole new world awakens and begins its "day." Countless numbers of insects and small beasts come forth from their various perches, nests, burrows, branches, and forts. Lemmings, shrews, rats, mice, rabbits, and even moles begin their nightly activities. The air is filled with nighthawks, whippoorwills, bats, and owls.

But the only sense that is curtailed is the visual one. Although perhaps *our* most important sense, it is often less important to the creatures of the wild. Of course, this is especially true when darkness falls. Smell, touch, hearing, even taste, and a few others we shall discuss later are perfectly usable at night. All these senses therefore are highly developed among the night creatures, who of course are looking for a meal. By the same token, they are good for someone else to eat. The game that is played in earnest during the day is thus repeated at night, though the rules may be slightly different. Weasels, skunks, and other carnivorous animals find this a good time for their hunting. So do foxes, wolves, and even some members of the cat family.

Unusually large eyes are often a sure sign of a nocturnal creature. Those of the owl are well known. Even more startling are those of a funny little monkeylike animal called the tarsier. He has large, hairless ears, and his eyes are positively enormous in relation to his small head. His life might be a nightmare of attacks by owls and other nocturnal predators were it not for his ability to see in the dark better than any of them. Without moving his body, the tarsier can turn his head around in a flash and see

Tarsier.

what's going on in the rear. If he doesn't like what he sees, he can jump ten times his own six-inch body length to another perch and safety.

Running from danger shares with concealment the honor of being the most popular of defense mechanisms. In the next chapter we shall learn some of the ways animals can run from trouble.

3. Run!

I RECENTLY read a science fiction short story in which a young lady on Earth was talking to her boyfriend on Mars. It was a very touching scene. There was only one problem, which the author either ignored or was ignorant of. Radio waves travel at the speed of light, or 186,000 miles per second. This is pretty fast. On the other hand, Mars is very far away from us. The point, of course, is that it would have taken up to twenty minutes for the young lady's "Hello darling" to get to the young man and another twenty minutes for his "Hi, sweetheart" to return.

Such a delay would, of course, have spelled disaster for the author's love scene; but an analogous delay has saved the life of many an agile creature being hunted. In other words, fast though a bullet may travel, some animals are so alert and so fast that they can actually get out of the way of a bullet—even after the gun is fired. This is sometimes called "ducking the flash." For example, Manly Hardy, a Canadian expert on otters, tells of firing a rifle at an otter who had poked his head out of a fishing hole in the ice. The bullet struck just six inches beyond and would have hit the otter dead center—if he had been there. But he had withdrawn his head before the bullet struck. Several species of waterbirds, such as loons, horned grebes, and bufflehead ducks, can also dodge shots.

There are several factors to be considered here. A rifle bullet takes about 1/20th of a second to travel 40 yards, a charge of shot from a shotgun perhaps 3/20ths of a second. A sprinter seems to be off at the crack of the gun. Actually it takes him about 1/10th of a second to get going, even though he is expectantly waiting for the signal.

The decision to do something is made in the brain. It takes some small amount of time to make the decision, and it takes additional time for the message to travel from the brain to the proper set of muscles. Nerve impulses in man travel at something like 100–200 feet per second, so it can actually take 1/20th–1/40th of a second for a message to travel from brain to foot. Allowing additional time for the muscle to react, we see why reactions cannot take place instantly. The record for drawing and shooting a pistol is .12 second, and here only the hand and arm must be moved. It has been shown that drivers normally take considerably more time, with something like ¾ to 1¼ seconds elapsing between the advent of trouble and brake shoe touching the brake drum. (In one second, a car moving 60 miles per hour will travel 88 feet.) Clearly a man could never "duck the flash."

But otters and game birds are much smaller than man. This gives them several advantages over larger creatures, at least in this respect. Because they are smaller, nerve impulses traveling through their bodies have shorter distances to travel, so reaction times are correspondingly faster. Also, smaller creatures actually "live" faster. Everything about them happens quicker. They breathe faster, their heart beat is much faster, their entire metabolism is faster; and the smaller the creature, the "faster" he lives. A horse's heart beats 40 times a minute; a man's 70; a cat's, 150; a hedgehog's, 300; and a mouse's, 650. All the other body processes take place correspondingly faster. The fast, birdlike movements seen in mice, squirrels, and birds are tied in with this fast rate of life.

Smaller creatures are more agile for another reason. They have less mass that must be shoved around by their muscles. It takes longer for an elephant to stop his charge than it does a lion. The mass—that is, the amount of material—in movement is a direct measure of the inertia of a body in motion. Newton told us long ago that a body in motion tends to remain in motion and that it tends to keep going in the same direction. He showed also that the impulse to keep going is directly tied in with how much mass there is in motion. The smaller creature, with much less mass, can stop or change direction much more easily than the larger one (all other things being equal, which they usually are not). Any of you who have tried to catch a rabbit, mouse, or other small wild creature know how fast and agile they are. Flies and other insects must be surprised or they will get away. The spiked forelegs of a praying mantis can shoot out and capture a fly in 1/50th of a second.

Some animals, rather than ducking danger, prefer to jump out of the way of trouble. We have already mentioned the remarkable jumping ability of the tarsier. A number of animals have taken to this form of defense, though in different ways. The grasshopper, for example, is built for jumping, so it is no surprise that he can outdo the tarsier, being able to leap some twenty times his own body length, or ten times his length in a vertical jump. This is a most effective defense, for it adds the element of surprise. All of us have had the experience of stalking some small creature and then starting violently when he has leaped or even moved unexpectedly.

When the grasshopper leaps, he tumbles head over heels and sometimes lands on his back. Because he is so light he lands without injury. The leap is purely defensive; his only objective is to get as far away from where he is as possible. The frog, another famous jumper, adds an objective—a target—to his leap. He seems to know that he is safer in the water and usually is found near lakes and ponds. When approached, he invariably leaps back into the water.

But the frog is not an especially intelligent creature. How does he "know" that he is safer in the water? It is easy to say that he knows it "instinctively." But many scientists do not like the word "instinct" and rarely use it. The problem is that by using the word, people are prone to believe that they have "explained" a behavior pattern, whereas they have really only appended a word to it.

With this in mind, researchers at the University of Oxford designed the following experiment. As shown in the drawing, a frog was placed in front of a two-opening frame. Each opening could be illuminated with light of different colors. After trying many different combinations, it was found that the frog's favorite color, by far, was blue; the frog almost invariably leaped toward that opening, particularly when provoked.

In other words, when frightened, the frog does not leap toward water but toward the color blue! Because water almost always has open sky above it, this preference

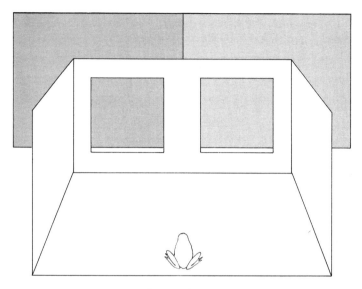

works just as well as if the frog had figured the whole thing out for himself.

Leaping is useful to large animals as well as small. The Rocky Mountain bighorn, a member of the sheep family, has taken up lodging in craggy wilds that are all but impassable to other creatures his size. Stories have been told of his being able to plunge fearlessly from precipices a hundred feet high, and even of his being able to leap up this high. This, of course, is nonsense. Actually he cannot jump more than six or seven feet high or more than about fifteen feet on the level. His reputation no doubt arises from the fact that he is fantastically sure-footed and sharp-eyed. He can leap from narrow ledge to narrow ledge with wonderful grace and apparent ease. His secret lies in tremendously powerful leg and foot muscles, plus two large, soft, rubber-like pads on each foot that can grip almost any surface, rough or smooth. Lack of footholds for the bighorn is an exceedingly rare occurrence, for gravity

33

prevents any large amount of rock from overhanging any distance. Virtually all cliffs lean backward to some degree and are provided here and there with ledges, bumps, shelves, and so on. If the bighorn can find two-inch footholds not more than five or six feet apart going up, or *twenty feet apart* going down, he has all the stairway he needs. Even the puma, his major enemy and a smart, amazingly agile creature, is not able to follow.

Another remarkable jumper uses his abilities in quite a different manner. The impala, a type of gazelle, lives on flat rather than mountainous territory in Africa. This species is found in herds numbering from ten to several hundred individuals. Each herd is led by an old male and often includes several "sentinels" whose sneezing alarm sends the whole troupe bounding away in soaring leaps ten feet high and covering distances of almost thirty feet. Once the impala is in action, lions, leopards, cheetahs, and hunting dogs sometimes give up right away. Generally the impala's predator depends upon surprising and catching one before he can spring into action.

Other uses for the impala's graceful and prodigious leaps have also been suggested. For one thing, he inhabits a type of countryside, light bush, where lurking enemies cannot easily be spotted—from the ground. A view from on high could be very useful. Equally logical is the suggestion that the leaps are also a fine warning signal if any impala missed the original alarm or if no warning was given.

Most wild hoofed animals, however, depend upon their speed, just plain old running, to get away. This defense, along with a good sense of smell and perhaps keen sight, makes the work of the predator very difficult. The Indian tigers, for example, must work hard for their food, roaming many miles nightly and making many unsuccessful attempts before successfully capturing a meal.

The speed of some animals is really surprising. The pronghorn, or North American antelope, is a lightly built animal standing only about three feet at the shoulder. With his sharp eyes, he can detect an object, especially a moving one, at a great distance across the open plains of his native environment. When startled and on the move, he can get up to fifty miles an hour, and one buck is said to have been timed at sixty miles per hour.

Although high speed is useful, it can also be dangerous. Deer have been known to impale themselves on projecting limbs while running at high speeds. Even if they see the danger, they might not be able to stop or dodge in time.

Blacktail doe runs at high speed.

Defense Mechanisms

Strangely enough, the fastest land animal in the world is not a preyed-upon creature but a predator. The cheetah, a member of the cat family, can travel at a speed of sixty-five or, in a short spurt, seventy miles per hour. But his endurance is low. Normally if he has not caught his prey after about a quarter of a mile, he slows down to await a more favorable opportunity.

When man began to hunt, many thousands of years ago, he added a new complication to the story. Two-legged man, though not nearly as fast as many of the four-legged creatures (he sprints at about twenty miles per hour), has amazing endurance. Even though the cheetah is far faster, he often loses his prey; the human hunter, however, can follow it for hours on end. Eventually the creature must rest, especially during the hot midday, for he does not have the efficient heat-dissipating mechanisms of man, such as sweating. The animal therefore simply falls down exhausted, providing an easy kill for the human predator. Some bushmen of the Kalahari desert in South Africa still pursue and capture large antelope in this way.

Because speed is not always a sure means of defense, some animals found another way out; they took to the air.

A good example is the flying fish, who eludes attackers by escaping from the medium that effectively imprisons his pursuers. Actually, the flying fish does not fly; he glides. Nevertheless, he sometimes manages to attain distances of as much as a fifth of a mile. He starts his flight by "taxiing," with his tail providing propulsion in the water. When his speed reaches some twenty-five to thirty-five miles per hour, he leaps out of the water and spreads his specially adapted fins. Sometimes rapid tail movements continue in the water even after the rest of his body has emerged, helping to build up speed. Because he cannot flap his "wings" and actually fly as birds do, the distance of his "flight" depends upon his initial speed, just as would be true of a hand-thrown glider.

Flying fish.

Defense Mechanisms

The squirrel family also has representatives who have "wings" to help them increase the distance of their already surprisingly long leaps. A point worth noting here is that the bushy tail of even the nonflying squirrel acts as a sort of parachute. The flying squirrel has gliding membranes. These are thin flaps of skin that are stretched out tight when all the animal's legs are fully extended, as they normally are in a leap. With his legs outstretched, he can sail through the air for long distances. Although he may guide himself somewhat by using his powerful tail as a rudder, he never tries to gain altitude by flapping his "wings." There are also flying lizards and flying lemurs although they, too, should more properly be called gliders.

There seems little doubt that gliding or parachuting preceded the development of true flight. However, it is not known whether the preliminary form was that of a creature like the flying lizard who used imperfectly developed wings to aid him in eluding enemies or whether it occurred in a ground-living form that flapped some kind of rudimentary wing to increase its speed of running, as does the ostrich today.

Whichever it is, flightless birds of today—including, in addition to the ostrich of Africa, the kiwi of New Zealand, the rhea of South America, and the emu and cassowary of Australia—are believed to be creatures that sometime in their history have lost their powers of flight. All live in areas that are, or at least were, devoid of predators, so they gave up their powers of flight. For flying *is* an effort and requires such very specialized characteristics as tremendously powerful chest muscles and a carina or "keel" to which they are attached, spindly legs, a small head

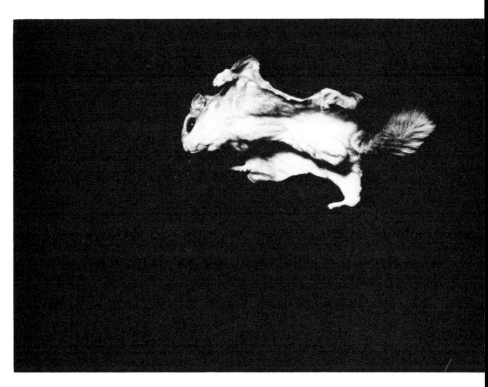

Flying squirrel glides to safety.

without teeth (to conserve weight), and light, hollow bones. Indeed, the wings of most birds are so fragile that such very light objects as twigs, branches, and even fishing rods can damage or break them.

It has been suggested therefore, that because flight was not necessary for defense, these birds simply gave it up, thus conserving their energies for the important job of obtaining food. Also, because there was no longer a weight restriction, the nonflying birds could begin to increase in size.

Flying birds, however, reach a weight limit that is far below the present-day ostrich's 300 pounds. This limitation exists because, as any textbook on aeronautics will tell you, the bigger a heavier-than-air machine is, the faster it must fly to keep itself up. The power requirements go up so rapidly that a point is quickly reached where the weight of the bird's flying muscles (the engines) begins to outstrip their ability to provide the necessary speed. As a result, one of the heaviest flying birds, the South African bustard, weighs no more than about forty pounds.

This of course is fortunate, else our ancestors might have had to contend with eagles the size of lions—and we might not have been here to discuss the problem.

4. Taking It

HERCULES was a hero of ancient Greek legend who was forced to perform twelve all-but-impossible labors. The second of these is pertinent to our discussion, for he had to slay the monstrous reptile Hydra, which had nine heads. As if this weren't bad enough the Hydra had additional defense mechanisms that under normal circumstances would be quite unbeatable. One of the heads was immortal, and no sooner was one of the other eight cut off than two grew in its place!

As you know, Hercules had great difficulties carrying out his second labor until, with the aid of his nephew Iolaus, who brought him a burning torch, he seared the neck as he cut off each of the eight heads so that it could not sprout again. The immortal head could not of course be slain. Hercules finally thought of a solution: he buried it under a huge rock. Perhaps it is still there, gnashing its teeth and threatening revenge.

A preposterous story? Perhaps. Yet it seems that no matter what man dreams up, nature has done it before him. We have already come across life that is "immortal" —the paramecium that never dies a natural death. As for the heads that grew back, this too has its counterpart in nature. Many creatures actually do have this very same ability to regenerate certain body parts that have been cut off.

When attacked, the "glass snake" sheds his tail. While the attacker gives his attention to the still wiggling tail, the glass snake (who is a lizard, not a snake) slithers quietly away, to grow a new one at his leisure.

The common earthworm can grow a new head as well as a new tail. Crustaceans, such as lobsters, crabs, and crayfish, add a neat trick to this kind of defense mechanism. They are equipped with strong pincers on the first pair of legs. If one of these creatures is in trouble but can get a firm grip on the attacker with one of the nippers, he clamps the pincer and at the same time sheds the whole device. He then escapes to safety while the attacker tries to pry himself loose from the nipper's grip. After a while the lost appendage grows back.

Although nature's creatures do not have the two-for-one arrangement of the mythical Hydra, some of them have another ability that is quite as effective. One species of worm, the longworm *Allolobophorus longa,* can be cut into as many as twenty pieces—most of which will grow into a new worm!

In one experiment a new head (and brain) was produced twenty-one times and a new tail forty-two times. Each successive growth, however, took a little longer, and finally the ability ceased altogether.

Regeneration is common in our experience. Plants are often pruned or cut back to shape them or promote thicker growth. Cuttings from certain plants can be replanted and will grow into complete plants.

Some of the lower animals, such as sponges, can perform even more remarkable feats. The sponge can be ground up into individual cells that continue to exist for several weeks exactly like protozoa. If certain types are squeezed through fine silk, a milky fluid results. If this fluid is put into a container of sea water, the cells will in time begin to clump. These are young sponges! After three months or so they will grow into mature sponges and will reproduce.

Another marine creature of particular interest to us stands like a small, delicate vase attached to the floor of lakes and ponds. It is roughly half an inch tall, has six to nine tentacles waving around from its "head," and is known for its powers of regeneration: it can be chopped

Hydra.

up and almost any piece will regenerate. It normally reproduces by budding, however, which means that "ba-
bies" simply grow out the side of the animal and drop off
when complete. This creature was first studied in the mid-
1700s by the Swiss zoologist Abraham Trembley, who
noticed that the buds may be quite numerous before
separating from the mother. This gives it a many-headed
appearance. Thus it has been named, quite appropriately,
the hydra.

As we go up the scale of life, the powers of regeneration
decrease. Salamanders can regenerate only (only!) a missing tail or leg. Some animals, such as frogs, can perform
this trick only in their younger days—as tadpoles. Higher
animals do, however, possess *some* powers of regeneration. The replacement of nails, skin, feathers, and antlers
is a kind of regeneration, as is the repair of broken bones
and torn skin.

Another way that organisms can protect themselves is
to have body parts or growths that are expendable. Some
insects, for example, have several hundred legs. Clearly
the loss of one or two will not matter at all.

Even an insect, with only six legs, can afford the temporary loss of one or two. These will subsequently grow
back, at least partially. Wings and antennae are also re-
growable.

The thick fur seen around some animals' necks—for example, the lion's mane—provides protection for his most
vulnerable part. After all, only a thin layer of skin protects
the large blood vessels, windpipe, and nerves that pass
through the neck. But the magnificent manes seen in zoos
are rarely seen in wild creatures, for thorns, thickets, and
other creatures often tear out great sections. The same

holds true for the ruffs of tigers, lynx, wolves, and rein-deer.

Even the hair on our own heads is a protection that we don't think about until we begin to lose it. A bald-headed man does get more scratches and bumps on his head than he did when he had a full head of hair.

The "dispensability" principle is used in other ways too. If you have ever tried to grasp a moth by the wings, you may very well have ended up with a batch of scales rather than the moth, for his body and wings are covered with tiny scales, which act like lubricating flakes of powdered graphite. Insect-eating monkeys often catch insects with their hands. It is likely that these dispensable scales are very useful in helping moths elude such attackers. But the most important application of the scales is undoubtedly in protecting moths against being caught in spider webs.

Suppose a moth collides with a web, which can range in size from a few inches to a few feet. Because the scales are only loosely attached to the moth, they are easily torn away and allow him to escape. Although the scales do not regrow, there are enough of them to serve in several en-counters with webs or various predators.

The porcupine has another kind of disposable protec-tion. Interspersed in his fur are many lightly attached spines or quills. These usually lie flat on his back but are raised when he is frightened or angry. An attacker with several of these barbed spines embedded in his skin will rapidly lose his urge to fight. Indeed most creatures who have had any experience with the porcupine will leave him alone, particularly if he has experienced a particu-larly nasty trick developed by the spined beast. Every long hair in the fur of the porcupine is attached to very

Porcupine.

sensitive nerve endings. When the porcupine is touched anywhere, he whips his tail around, and the attacker is hit with a batch of sharp, stinging quills. The sensitivity of his hairs explains how he can accurately swing his tail even when his head is buried or his eyes are closed. Another approach is to suddenly run backward into the attacker, which usually prevents any further attacks.

Spines are a widespread means of defense. Surgeonfish have knifelike blades located in pockets on each side of the body just in front of the tail. When these herbivorous fish are in danger, the blades spring out and are used for defense. They are sharp enough to produce a nasty cut.

Triggerfish also utilize a spine, but this is on top of the fish and they add to it a locking device. The first and second spines on the back have an interlocking device. When the first spine, which is the large one, is raised, it is held firmly in place by the second one and will remain in the erect position until the second spine is released. The tall spine serves at least two purposes. It makes being swallowed by a larger fish very unlikely. But, in addition, these fish are often chased into corals by attacking fish. The triggerfish can then raise the large spine and anchor himself in this way.

One of the best-protected creatures around is the sea urchin, a hollow, bony ball bristling all over with spines. He has no unprotected underbelly, so he presents a formidable barrier against attack. This relative of the starfish has no other means of defense—and needs none. The land-living, quilled hedgehog overcomes the vulnerability of his unprotected belly by curling into a ball with his nose tucked between his hind legs.

Many plants, too, particularly cacti, are protected by spines and thorns, perhaps because they cannot afford to lose moisture in their dry desert homes. The spines act as protection against hungry and thirsty animals. It is also worth noting that plants lose most of their water from leaves. Thus many desert plants appear to be all stem, with their leaves reduced to the very useful spines.

This kind of protection is unusual in trees. One exception is the Central American acacia (*Acacia sphaerocephala*), whose trunk and branches are covered with large, sharp thorns. As if this weren't enough, the thorns are hollow and are likely to be the home of stinging and biting ants, which swarm out when the thorn is touched.

Defense Mechanisms

The porcupine fish also has spines but has carried his defense to even greater lengths. Indeed, he is probably the best-protected fish of all, for when menaced, he can swell up like a balloon. This makes it even more unlikely that he will be swallowed accidentally by some near-sighted fish with a big mouth.

Other fishes, such as the puffer, globefish, and swellfish, can also blow themselves up like a balloon. The puffer is an eight-inch fish and can swallow a quart of water. As an alternative, he can swallow air and float on the surface of the water, belly up. Inflated, he feels something like a basketball. Should a larger fish decide to go ahead anyway, he would be in further trouble, for the puffer is poisonous.

The puffing up serves at least two functions. First, it makes the fish hard to swallow, and second, it makes him appear larger and more of a match for the predator.

Perhaps the next best thing to having a coat of spines is

Inflated swellfish.

to carry about a suit of armor. Indeed, long ago, before the age of reptiles, the most popular apparel for fishes was a stiff set of bony scales. Later, armor was widely used by humans but lost its usefulness with the advent of firearms in the Middle Ages. In a similar manner, the development of large, bone-crushing, fish-swallowing monsters began to put the armored fish out of business. Speed and other defense mechanisms became the order of the day.

Few modern fish are armored. An exception is the trunk-fish of tropical waters. The trunkfish's body consists of a series of hard plates joined together to form a solid bony case. A tortoise can at least move his head around when he wants to. In the trunkfish only the jaws, eyes, fins, and tail are free to move. But he's still around, so apparently the arrangement works.

Among armored land animals one of the best known is the armadillo, whose upper parts, even the ears, are covered with sculptured bony plates. Two of them, one cov-

Armadillo.

ering the rear end and the other the shoulders, are quite large. Between these two there are bands or joints that permit the armadillo to curl up and protect himself as the hedgehog does. The tortoise, of course, relies on pulling his head and limbs into his fortress for protection. This mode of protection seems to work very well. Several tortoises have been reported to be 150 to 200 years old!

One small species of armadillo (the word derives from the Spanish term *armado*, "armed") lives entirely underground, like the mole. This one is called pichiciago (*Chlamyphorus truncatus*) and ends in a flat bony shield, as if his rear end had been sheared off. The advantage here is that if chased, he can squeeze into his burrow, thus presenting to the attacker only a hard, relatively impregnable plate and nothing to grab.

Unfortunately for tortoises, their shells are very attractive to man, who has long used them for ornamentation. Shellfish too provide many beautiful specimens and are avidly collected by hobbyists. Shells are produced as hard, limy outer skeletons by the vast majority of the mollusks, a group that includes clams, snails, and oysters. The shell material emerges as a liquid secretion from an organ called the mantle, and then hardens. A small amount of the material, calcium carbonate, is produced by the mantle for repairing cracks and holes and also for encasing foreign particles that might enter the shell and cannot be digested. In the oyster this latter process produces the very beautiful pearl.

Oysters and clams are bivalves; that is, their shells take the form of two halves, hinged at the top and controlled by powerful muscles. Most of them can move in a vertical position along the sea or lake bottom by means of a "foot," which is protruded from the slightly opened shell. Swimming ability is rare, although the scallop manages this by clapping his two shells together, which sends him on trips of a dozen or two feet at a time. One clam, *Tridacna gigas*, is protected by what is probably the ultimate in armor: a large, thick, virtually impenetrable shell. This giant, who lives on the coral reefs off the coasts of the various islands of Indonesia can reach four feet in length and several hundred pounds in weight. Oddly, his food is almost microscopic in size. Although stories have been told of natives who were caught in the viselike grip of the shell, these have not been authenticated.

The whole point of armor, of course, is to protect the life of the wearer. This is seen in its purest form in an unexpected place, namely, seeds. The tenaciousness of life is best seen here, too. Indeed, it is sometimes beyond belief, for some seeds can be boiled or frozen without harming them. Some resist continual boiling for forty-eight hours. In one experiment, alfalfa, mustard, and wheat seeds were perforated, dried for six months, sealed in a vacuum for a year, and then placed in a below-zero freezer. And still they germinated (sprouted). Weed seeds have been buried in glass bottles for forty years. They were then dug up—twenty years after the first experimenter had died—and planted. Half of them were still alive. Indian lotus seeds have been discovered in a Manchurian lake bed that radioactive carbon dating showed to be about 1,000 years old. Some of these too still had life in them.

Certain animals may act in ways that are no less surprising. They will appear to be dead when they really are not. Indeed, the trick of "playing 'possum" is quite widespread in the animal kingdom, being practiced by many insects, spiders, and centipedes, as well as certain snakes and birds. The expression derives from the characteristic of the opossum of not providing the slightest resistance to enemies. He will often half close his eyes and simply hang his head and tail. If pushed he will fall over, apparently dead. This may puzzle the predator enough to get him off without injury.

In this chapter we have seen some of the ways in which animals and plants "take it." We turn now to a somewhat more "offensive" type of defense.

5. Chemical Warfare

THERE are few places on Earth where life of some kind has not taken hold. Flowers have been found in the frozen wastes of the far north and penguins in the far south; bacteria manage to survive in boiling hot springs; and both plants and animals exist in supposedly barren stretches of hot, dry desert.

The most extreme desert conditions in the United States are found in Death Valley, California. Although the area is only 200 miles from the Pacific Ocean, the tall Sierra Nevada provides an effective screen against that source of water.

As a result, the average total yearly rainfall in Death Valley is about an inch and a half—as much as other sections of the country get in a single heavy downpour. There is almost no surface water, and only a few springs bring up the small amount of runoff from surrounding mountains. Yet in this arid land can be found patches of lush greenery, such as the evergreen creosote bush.

How does the creosote manage to survive? First of all, its roots reach so far out that it can extract water from a large volume of soil. But even more interesting is the fact that two such bushes never seem to grow near one another. It is not at all surprising in other areas to see a

53

small maple taking root near a grown tree. Clearly if this happened with the creosote, the bush would be in serious trouble. There is little enough moisture as it is.

Obviously the spreading out is not a conscious activity. What happens is that the roots of the bush excrete poisonous substances that kill any seedlings that start near it! The system works very nicely and is correlated logically with the amount of rainfall. The more rain, the closer the bushes grow to one another. Apparently, rain leaches or soaks out some of the poison from the soil so that it does not contaminate so wide an area. Thus the less rain there is, the larger the clear area around each bush.

A very similar phenomenon was seen about forty years ago—one that was to have world-wide consequences. The Scottish-English bacteriologist Alexander Fleming had left exposed in the open air some culture plates of bacteria with which he was going to do some experiments. Returning to his laboratory a few days later, he found that a speck of mold similar to that seen on old oranges had fallen on one of the culture plates. He was about to discard the plate, because the culture was no longer pure, when he noticed a clear area around the speck of mold. This indicated that some unknown substance in the mold had killed the bacteria or at least prevented it from multiplying.

The mold was the fungus *Penicillium notatum*, which, I am sure you have guessed by now, yielded the wonder drug penicillin. The point, of course, is that penicillin is poisonous, but to certain disease-causing organisms, not to man. It's all a matter of perspective.

Scores of other substances with similar life-saving properties have now been found; some of them must be

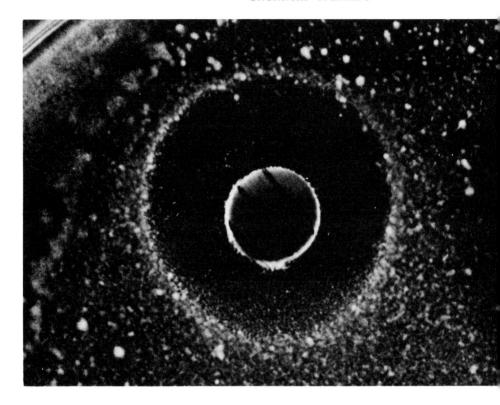

Penicillin disc kills bacteria, creating clear zone around it.

handled very carefully because they can be toxic to man also. These drugs are generally called antibiotics, which literally means "antilife." What we have then is a method of putting one microscopic life form against another.

An antibiotic is a substance produced by a living organism that can check bacteria or other disease-producing organisms. Although some antiobiotics actually kill bacteria, most do not. Rather, they inhibit the growth, multiplication, or other life processes of the invading germ. The pleasant result is that the balance between victim and invader is altered in favor of the host (us), and our basic defensive system is then able to destroy the invaders. Unfortunately, none of the new drugs works against viruses, which is why there is still no cure for that most common disease, the cold.

Enemy microorganisms do their damage in various ways. They may destroy body cells directly, or they may deprive the cells of essential nutrition. Microbes may also release in the body products of their own living processes that are toxic (poisonous). The toxin produced by the diphtheria bacillus, for instance, has a chemical resemblance to snake venom! If not met with by some artificial antidote, the body's defensive system must handle the job, if it can.

Of course, part of the body's defense program is to prevent virulent microorganisms from entering the body in the first place. Thus, many parts of your body, which you would not normally connect with protection against disease, are truly part of your defensive system. Indeed, there is a whole set of outer fortifications. That outer covering called your skin is actually the first line of defense. It is an elastic yet tough and effective suit of armor. As long as it is uninjured, no microbes can penetrate. If cut or bruised, it will mend itself. In the interim, however, there is danger, which is why antiseptics are necessary on the wound.

As with any kind of armor, there are openings in the skin through which invaders can enter, namely, the openings through which we take in food, oxygen, sounds, and those through which we discharge our wastes. At all of these openings there are safeguards. Tears and saliva, for example, contain a microbe killer called *lysozyme*. Although the ancients knew nothing of bacteria or lysozyme, they did realize that saliva had curative powers, simply by observing that animals lick their wounds and the wounds usually heal.

The nasal opening is guarded by little hairs and farther on by sticky secretions of the mucous membrane. This is a type of skin covering found in almost all the passages that connect the internal and external portions of our bodies. Along all the linings from nose and mouth to the lungs and along the digestive tract a slimy coating traps bacteria. In the upper lung passages the cells have very fine fingerlike projections, called *cilia*. These "beat" steadily and create a current in the mucous, thus pushing it and the bacteria out into the throat. This is why you must be so careful when coughing or sneezing (so as not to infect others around you), particularly when you have a cold or other respiratory illness.

When microorganisms do get into the digestive tract, many are killed by the digestive juices. If they survive these, they are pushed through the canal by waves of contractions (peristalsis) along passages whose mucous lining gives them (hopefully) no place to catch hold.

When this doesn't work, vomiting or diarrhea may result. Although unpleasant, these processes help rid the body of undesirable visitors. But sometimes, in spite of all these defenses, germs manage to slip through into our tissues. Then a new series of defenses goes into action. The redness, swelling, and fever associated with an infection are the work not only of the invader but also of the body's inner defenses. As we know, microorganisms multiply rapidly. The rescue forces therefore must also act rapidly. Body fluids rush to the scene. These, particularly the blood at first, bring an army of white cells, whose major function is to do battle with the enemy.

When all is well, there is a fairly constant proportion of red and white cells throughout the body (about 500 red to each white cell). All circulate constantly throughout the organs and tissues. When a foreign substance breaks through the outer fortifications and begins to damage the body's cells, histamine and other chemicals are released. These cause dilation (enlargement or stretching) of the blood vessels and large numbers of the white cells work their way out into the surrounding tissue.

As these cells rush to the scene, often leading to swelling, they begin to perform a double action. Some of the white cells surround the infected area (most microbes concentrate in certain portions of the body). They actually "quarantine" it, making a wall with their own bodies. Within the barricade another type of white cell sets upon the enemy and literally eats it up—in very much the same way as the single-cell amoeba engulfs and digests its food particles.

The battle is on. How are white cells guided toward an infected region? Every such region contains numerous dead cells plus the substances into which they decompose. Thus waste materials stream away from a site of infection, and white cells, moving by, are stimulated to move toward the site. It is a purely chemical reaction (though a complicated one to be sure) that is similar to what happens when an amoeba moves toward a food source or away from a poison.

Much depends upon whether the defenders can gobble up the invaders faster than they can multiply. If not, all is not lost, for the body can then fall back on yet another set of defenses.

Many of the dangerous substances that invade the body are proteins. Examples are the virus itself or the products of bacteria, plant pollen, or other protein-containing organisms. These are called *antigens*; they stimulate the body to produce antibodies, which are complex protein molecules that combine with the antigens in a kind of "lock-and-key" reaction. The antibodies are specifically constructed to match or fit the molecular shape of the antigen. Thus an antibody for one disease will probably be useless for another. However, once the body has manufactured a set of antibodies for a particular disease, it will "remember" the pattern, and so the body is *immune* to that disease for varying periods of time, perhaps for life. The body can be stimulated to produce such antibodies by an injection of weakened or dead microorganisms (vaccination) that still retain their characteristic "antigenness." The trick, of course, is to present the proper pattern, or template, without actually infecting the body. It was by this method that the dreaded disease poliomyelitis (infantile paralysis) has recently been attacked and virtually wiped out in the United States. In this case, it was found that shots were not necessary; the vaccine could be taken orally.

Among the organs that produce antibodies are the lymph nodes and the liver. Lymph nodes are pea-size structures located strategically in various parts of the body (for example, the neck and armpits). When called into action these may become swollen and sore: we have "swollen glands." The cells of these organs apparently absorb samples of invading antigens and then manufacture large quantities of the appropriate antibodies. This is why it takes time for this line of defense to come

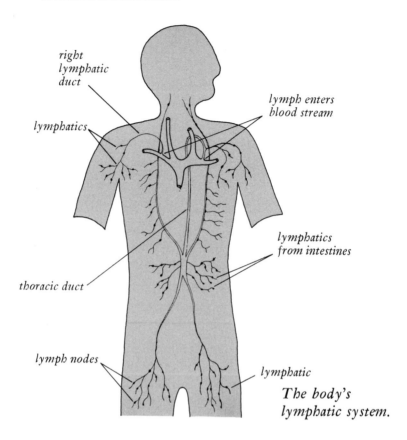

right
lymphatic
duct

lymphatics

thoracic duct

lymph nodes

lymph enters
blood stream

lymphatics
from intestines

lymphatic

*The body's
lymphatic system.*

into play. Antibodies work in several ways. Some pathogenic (disease-causing) bacteria have an external layer of slime that helps protect them against ingestion by white blood cells. Antibodies may coat the bacteria so that a white blood cell can get a grip on it. Antibodies may also cause a number of tiny viruses to stick together so that they cannot enter and infect a cell.

The various sophisticated chemical defense mechanisms we have discussed so far developed rather late in the history of animal life. Antibodies, for example, are found only in vertebrates (animals with backbones). A recent discovery disclosed that these defense mechanisms are backed up by what may be a far more primitive one —indeed, one that even plants may possess in simplified form. Our earlier discussion of the antisocial activities of the creosote bush and penicillin gives us a clue. This mechanism turns out to be a chemical that goes into action long before antibodies can be manufactured. Unlike antibodies, it can be manufactured by all the ordinary cells of the body. It is a protein that has been named *interferon*, for it has been found to interfere with the reproduction of viruses. Moreover, it does this inside the cells, a refuge taken by viruses that antibodies have difficulty reaching.

Unfortunately, the most dangerous viruses seem to provoke only a meager production of interferon, which may be why they are so dangerous. Hope is raised by the fact that interferon, when present in adequate quantity, can successfully cope with many viruses. New techniques are being developed for handling some of the still troublesome viral diseases (influenza, measles and some think, even certain forms of cancer)—for example, by injections of interferon or of other chemicals that would boost its production.

Fever, interestingly, seems to boost the body's production of interferon while at the same time being harmful to viruses. Thus, at least in some instances, fever may be a factor in controlling disease as well as a symptom of it.

It is interesting to note, too, that certain diseases and allergies are thought to be the result of the body's defense system going haywire. In allergies, for example, instead of a person becoming immune to an antigen, he becomes hypersensitive or oversensitive, to it. When a person suffers from hay fever, his body produces a special kind of antibody that attaches itself to cells containing histamine. If a bit of pollen lands somewhere in the respiratory system, it produces an antigen that interlocks with the special antibody. The complex molecule that results triggers the cells to release their histamine. The result is a chain reaction of response that is supposed to come into play only in the face of serious infections. The blood vessels become dilated, fluids are produced and seep out of cells, and nose and eyes begin to run. The sufferer is the victim of his own antibodies.

No one is sure whether the poisonous and irritating substances in plants are simply waste substances or whether they are a means developed somewhere in their history as a defensive adaptation. In either case, however, these chemicals do protect the plants against "attacks" by insects, animals, and man. It has been found that even many of the more common trees emit potentially dangerous substances known as terpenes and esters into the air. Many of these compounds—the terpenes, for example—are of no obvious use to the plants that produce them. However, some have been found to repel insects that might otherwise feed on the foliage. And, of course, the substance that varnishes the leaves of the poison ivy plant does a very good job of keeping away a large group of animals, namely, human beings.

The interesting part is that no matter how virulent the poison, some creatures seem to be immune to it. Not every person who walks through a poison ivy patch is infected. And the fly amanita mushroom *Amanita muscaria* is deadly poisonous to man but is harmless to red squirrels and chipmunks.

Many animals, too, use chemicals as a defensive measure. These take many forms. The odoriferous emission of the skunk is only the best-known one. A number of small animals, such as the weasel and civet cat, can perform the same trick, as can certain beetles and caterpillars. The caterpillar of one moth (*Notodonta concinnula*) emits hydrochloric acid when disturbed, and another (*Dicranura vinula*) has large amounts of formic acid in its spray. The bombardier beetle can direct its spray with remarkable accuracy at distances as great as twenty inches.

A slightly different approach is taken by other creatures. The skin of the wild pig, for example, exudes an unpleasant-smelling substance when that animal is in danger, and some frogs and toads are equipped with warts on their bodies from which a repulsive slime oozes. In some cases, as in the American pickerel fish, the skin secretion is so powerful that it will burn the skin of any human who touches the creature.

Another way to avoid being eaten is to *be* poisonous. A number of fishes fall into this category. One of the most deadly is the puffer, which has been known to cause death in twenty minutes. But, as has happened so many times, man has gone against the expectations of nature. The Japanese have developed a special way of preparing the puffer so that it is edible. Indeed, the puffer is considered a great delicacy. If the fish is not prepared in the proper manner, however, it remains poisonous!

When a creature has a means of injecting or otherwise transmitting its poison to another one, it is said to be venomous. Certain venomous fishes have poisonous spines or barbs of one kind or another. Some fishes of the Indian and Pacific Oceans can cause agonizing death. One of the most deadly, the stonefish, unfortunately looks like a rock and is therefore easily stepped on by mistake. The result is an injection of deadly poison into the foot from one of the spines on his back.

Certain caterpillars have a similar means of defending themselves. Contact with certain stinging caterpillars has been known to make hospitalization necessary. There are about fifty species of caterpillars that are venomous, most of which are either strongly colored or strange looking in some way. However, one of the most frightening looking of all, the many-spined five-inch hickory horned devil, is not venomous. We shall see later that there may be a reason for this.

Stonefish—easily mistaken for a stone.

6. Fight!

ALTHOUGH most animals that are hunted by others would prefer not to fight, many will do so if necessary. Ostriches, for example, rely primarily on their remarkably high speed for defense, but they are formidable fighters when cornered. Using their almost-hooflike feet, they can deliver vicious blows that will deter all but the most determined enemies. Furthermore, their sharp claws can cut like knives.

Zebras, wild horses, and giraffes can also do considerable damage with their tough hoofs. If not immediately overwhelmed, the zebra is quite capable of hurting even as powerful an animal as a lion.

Members of the cat family (felids) generally attack from the side or rear, their killing bite often directed to the back of the neck, but lions, leopards, and particularly the cheetah also kill large prey by gripping the throat until the animal suffocates. Cats have long claws and powerful forelimbs, which they use to fend off the defensive weapons of their victims. Members of the dog family —wolves, foxes, and wild dogs—are not so equipped and generally attack from the rear. Although not quite so well adapted for a carnivorous existence as the cats, these animals (canids) manage very well. With their fine sense of

Defense Mechanisms

smell, long legs, and deep, narrow chest, they can seek out and run down the fastest prey through superior endurance. Jackals and foxes, however, usually sniff out and pounce on their prey rather than run it down.

In a knock-down, drag-out fight animals will use teeth, hoofs, claws, nails, horns—whatever they have as fighting equipment. (There are exceptions, which we'll discuss later.) The giant front teeth of the baboon may very well save him from a smaller leopard.

The teeth of animals with backbones (vertebrates) are all similarly constructed. There is an inner mass of soft, pulpy material containing blood vessels and nerves. Surrounding the pulp and making up the bulk of the tooth is a hard, bony substance called dentine. And, finally, the

Baboon and his giant teeth.

exposed portion of the tooth (the part above the gum line) is capped with an additional layer of enamel, the hardest substance in the body.

Ordinarily teeth cease growing after reaching a specific size. Among gnawing rodents such as rats, beavers, and squirrels, however, the teeth continue to grow to allow replacement of the curved, powerful incisor teeth as fast as they are worn away.

Occasionally one of these teeth is lost in some way or is deformed so that it no longer meets its opposite number in the other jaw. If this occurs in the lower jaw, the unopposed upper tooth may continue to grow in a circle. Should this continue long enough, the tooth may reenter the roof of the mouth from below and perhaps even kill the animal by slowly penetrating his brain.

The tusks of the boar and elephant and the spear of the unusual narwhal whale are similarly constructed of unopposed, continually growing "teeth." The narwhal is a ten- to sixteen-foot whale that inhabits the far northern seas. Its tusk, elegantly sculptured in a tapering spiral, can measure anywhere from four to nine feet in length and typically grows out horizontally through the upper left lip of the male. Sometimes tusks develop on both sides but never the right alone, and they are always spiraled to the left, even when the animal has two tusks. Stories are told of narwhals taking strategic measure against killer whales by forming a tight ring with the tusks pointed outward. The tusk may also be used for flushing schools of hiding fish, but no report of the tusk actually being used to spear another creature has ever been made. The tusks are apparently a sort of male sexual symbol, like the great feathers of the peacock.

Defense Mechanisms

Horns and antlers, although sharing some of the characteristics of tusks, are a different matter altogether—indeed two different matters. The horns of antelopes, cattle, and goats are made of the same material as hoofs, hair, and fingernails and toenails; that is, they are a hardened outer layer of skin. The horn grows over a bony core that is part of the skull and is a permanent fixture like tusks. After the animal dies, the horn is easily removed from the bony core. The old-fashioned powder horn used by frontiersmen is a cow's horn obtained this way. In most cases, both male and female of the species carry horns.

Rhinoceros horns are slightly different; they are more fibrous in nature, rather like masses of hairs cemented together, and lack the bony core. They are tough nevertheless and make formidable weapons, often growing to eighteen inches or more in length. As with tusks, they continue to grow. Being softer, however, they must be ground down to keep a point by goring the ground and by rubbing on termite mounds and trees. This also stops the horns from growing to uncontrollable lengths, which would make them useless for their primary function, defense. We can be quite certain that this is their primary object, because rhinos are vegetarians.

The black rhino of central and southern Africa has a greatly exaggerated reputation for being bad-tempered. As with most animals, he would prefer to be left alone and will normally attack only when baited or annoyed by humans in their noisy jeeps or Land Rovers. Adult rhinos are large enough and, with their formidable horns, are well-enough equipped to be left alone by most natural predators, although lions and crocodiles may go after young ones.

Two examples of horns: antelope (above) and rhinoceros.

Defense Mechanisms

The "horns" of deer, on the other hand, are true bone and are called antlers. They are shed and regrown periodically, normally once a year. Except for reindeer and caribou, they are carried only by the male.

The growth of antlers is one of the wonders of nature. Around late fall or early winter the antlers break off from their pedicles, or sockets, which are part of the skull. Where they break off a scar forms, which is quickly covered over with skin. After a short time small knobs appear that are very tender; the animal is very careful not to injure them. Within these knobs a bony structure forms; it grows rapidly, and soon the antlers take shape. For large deer, such as moose and wapiti, the growth continues for months, but by late summer or early fall the center of the antlers hardens. The antlers of a large moose have reached a spread of seventy-eight inches across!

On the other hand, we have the caterpillar of the

Antlers of a whitetail buck.

Harpya-spinner, which is related to the silkworm. It has two small horns, not on its head but on its rear end. If pursued by ants or other creatures, it turns and "runs." If the predator approaches to within an inch and a quarter, two whiplike extensions spring out of the horns and lash out wildly in all directions. Most enemies give up after being hit once or twice.

The scorpion, a relative of the spider, though larger, also uses his rear end, but in quite a different manner. Scorpions are flattened, segmented creatures having a poisonous sting at the back end. Normally this rear portion (which is part of the abdomen, not a tail) is curved over the back so that the spinelike, venom-bearing sting is in position to strike left and right and even in front. The scorpion's food consists of insects, spiders, and other small members of the animal world. But the sting is also obviously a very useful defensive tool, for scorpions are believed to be the oldest living land animals, having been traced back some 400 million years. (One of the few serious competitors is the cockroach, who has been around for some 250 million years. The cockroach depends upon other defensive measures, which we'll get to subsequently.)

Adult scorpions range from one-half to eight inches long and include some species with an especially potent venom. Size and condition of the victim are extremely important here, as well as with other poisonous and venomous creatures. The sting of the scorpion, for example, is not usually fatal to larger animals but quickly paralyzes the small creatures on which it feeds. However, the sting of certain scorpions can be fatal to small animals, children, and even infirm adults.

Defense Mechanisms

Indeed, of all the fighting techniques used by small creatures, there seems little doubt that the use of stings is the most effective one. The sting of bees, wasps, and certain ants provides these insects with a formidable defense (and offense) against their enemies.

In the bee and certain wasps the sting is associated with the egg-laying organ of the female. The male therefore is stingless. In the honeybee the sting is a minute needle with tiny barbed edges. This makes it difficult for the insect to pull the sting loose and often results in its loss and the subsequent death of the insect. Yellow jackets, hornets, and other wasps have sharp, smooth stings that can be used repeatedly. Human deaths occasionally result from a single stinging by a bee or wasp. This kind of hypersensitivity may be the result of an unusually acute allergic reaction.

The scorpion spider, *Mastigoproctus giganteus*, defends himself in a different way. Instead of a poisonous sting, he has acid glands that can spray a strong poison at enemies up to sixteen inches away. If menaced by a thick-skinned or armored foe such as an armadillo, he does not bother to spray the animal. Instead he sprays his own body. No sooner does the enemy touch him with nose or mouth than a stabbing pain causes a hasty retreat.

There seems little question, though, that in the realm of poisonous and venomous creatures, snakes present the greatest danger. In India, some 16,000 people die every year from the bites of cobras and other venomous snakes. In the Americas, rattlesnakes, water moccasins, copperheads, and fer-de-lances present the greatest danger. The venom from these snakes, all of which are called pit vipers, is driven into the skin by two fangs at the front of

72

the upper jaw. The fangs normally lie flat but can be erected at will.

The venom of the pit vipers and some of the spiders seems to be far more powerful than is required for ordinary food acquisition. The very nature of snakes' eating habits (swallowing creatures whole) limits them to small prey. Clearly, superpowerful toxins are not needed. However, they are useful in defense against a larger attacker. A well-fed rattlesnake, for example, will strike chiefly when pressed or perhaps stepped upon by accident. That it rattles a warning when approached and glides away when given half a chance emphasizes the defensive role of this potent weapon. We might compare this to the notice posted on trucks and houses to the effect that burglar alarms are installed. At first thought it seems silly to warn the burglar, but the main objective is to keep burglars away, not to catch them.

Of course, many creatures have more than one means of defense. An interesting example is the caterpillar of the tiger swallowtail butterfly (*Papilio glaucus*). First, he has two large eyes that he presents to an attacker. They are "painted" on his body and are large enough to represent a much larger creature than himself. His true eyes are much smaller and are located farther forward, on his head.

Second, he has the ability to give off a very disagreeable odor. Apparently, he feels relatively secure, for when menaced, he will not run but will rock back and forth,

Caterpillar—"painted" eyes showing clearly.

almost like a fighter, apparently "threatening" the enemy. After a winter of this rather unpleasant way of life, by means of which he can survive, he turns into the lovely tiger swallowtail butterfly.

One of the most unusual, and potent, defense mechanisms among larger creatures is the use of electricity. Although many species of fish can produce electricity, the electric eel (*Electrophorus*) is best known, and with good reason. This inhabitant of the Amazon and other South American rivers can produce up to 600 volts of electricity, enough to kill a large fish. A human in good condition would not be killed but would receive a tremendous shock and would be sure to leave the eel in peace. The importance of the electric organs to their owner is seen in the extent of their volume; they take up fully 40 percent of the animal's bulk and extend about four-fifths of the way back from just behind the head right to the tail.

The eel, and other fish similarly equipped, use the electricity both for obtaining food and for defense, for they have no teeth or other protective devices. Again, the defensive aspect is emphasized by the power of the weapon, for their food is typically quite small.

When the electric fishes were studied more carefully, it was found that certain species had very weak electrical powers—in the area of tenths of a volt. For a long time these powers seemed to have no functional value, and the organs producing them were called "pseudoelectric" ("pseudo" means false, make-believe). It was later found, however, that these weak electrical discharges probably serve several purposes, including detection of both prey and attackers.

These fishes are highly sensitive to changes in the electric fields they themselves produce. Various types and sizes of objects in the fields will distort them in characteristic ways, which the fishes are apparently able to distinguish.

The three-foot electric catfish lives in African waters and can deliver shocks up to 200 volts. He hunts only at night yet has very weak eyes. Although the skin of many creatures performs a variety of sensory functions such as providing sensitivity to touch, pain, and heat, the skin of this African fish has added the function of position finding. That is, the fish's surroundings distort his electric field in characteristic ways, which he has learned to "read" and use for purposes of navigation.

As we shall see in the next chapter, this is but one of a number of strange phenomena in the world of the senses.

7. The Senses

Let us begin this chapter with a simple though very revealing experiment. Set up three bowls of water—one hot, one lukewarm, and the last cold, in that order. Now put your left hand in the hot water and your right hand in the cold. Hold them there for a minute or so. Even with your eyes closed there is no question which is the hot and which the cold.

Now plunge both of your hands into the lukewarm water. You may be surprised to find that the water feels cold to your left hand and hot to your right one.

For an explanation, we return to the first part of the experiment. When you first put your right hand in the cold water, for example, there is a strong sensation of cold; but this begins to subside very shortly thereafter. Toward the end of the minute, the water begins to feel quite lukewarm. One tends to think that the water has warmed up. Certainly it has, but not nearly as much as it seems to. The point is that the cold and heat receptors in your skin respond to *change* in temperature, not to the temperature itself.

True, you can certainly tell the difference between a hot and a cold day. But this difference comes to you in many different ways. And except for your initial exposure to the heat or cold, your temperature sense organs do not participate. On a hot day you have the moist sensation of perspiration as well as sensations arising from automatic expansion of blood vessels and the consequent rapid flow of blood in your skin. On a cold day your major sensation, again after the initial shock, may only be the physical one of shivering.

The heat or cold is called a *stimulus*; the cells in your skin that respond to the stimulus are called sense organs or sensory cells. The fact that the temperature-sensing cells respond only to change of stimulus is not unusual. Indeed, this is, in general, true of all your senses.

Consider your sense of touch, for example. Take a deep breath and hold it. Now don't move. Think of the part of your skin under one of the sleeves of your shirt or blouse. The material is resting on your skin, yet you are not aware of it. But touch the material with your other hand; the slightest pressure will evoke a reponse in your touch-sensitive cells.

The sensitivity of these cells is increased manyfold by the hairs found on many parts of the skin. The fine hairs on your arm, for example, need be touched only so lightly for you to be aware of action in the area. Even a breeze can do it.

A hair does not fit tightly in the skin. Rather, it lies in a sort of socket, called a follicle. As shown in the drawing, there is a little belt of branching nerve fibers that surround the hair near where it opens to the outside world. Because the hair is loose in the follicle, the slightest movement of the hair is enough to activate the nerve and send an impulse coursing to the nerve centers in your body. In other words, the principle of leverage is being used.

77

Defense Mechanisms

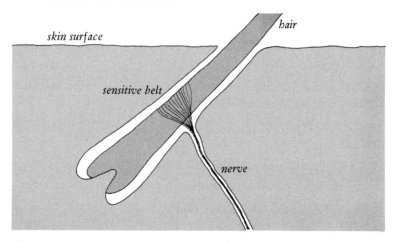

Slightest movement of hair is sensed by nerve net.

When you are high-strung, nervous, or frightened for some reason, you know very well that a simple touch on the hairs of your arm can cause you to jump several feet. Obviously, it is the *change* in the situation that did it. Once you realize that it was your friend or a projecting limb that was the cause, you quickly forget about that stimulus even if the pressure remains.

The usefulness of sensitivity to change becomes even clearer when we think purely in terms of self-defense. Consider a mouse embarking on a hazardous journey through the bedroom and to the kitchen cupboard for something to eat. He must be on the alert for enemies. When he first spies someone sleeping, his eyes must signal this, but they must not continue to hog the floor with their message, "That thing is motionless. That thing is motionless. That thing is motionless . . ." Rather, they must report essentials, which in this case is clearly anything that is moving. The slightest movement somewhere off in a dim corner has infinitely more interest to the mouse than a whole houseful of furniture and sleeping humans. And his eyes are equipped to handle the situation.

Human eyes show this same characteristic, for at the extreme ends of our field of view lie cells that respond only to movement. Indeed, we can see something moving way over at the side long before we can see what it is or even distinguish its color. Try staring at an object directly in front of you. Bring your arm way out to the side, behind your field of vision, and start to move it toward the center. Wave your thumb as you do it, and you will see movement before you can distinguish the outline of your hand.

Oddly enough, pain serves the same purpose—alerting us to danger—although we normally don't think of it in this way. For example, we need only consider the fix those people are in who *don't* feel pain. Consider a toothache, for example. This is never a pleasant experience. But if teeth didn't have nerves in them to signal when the hard

*Mouse—all senses
on the alert.*

outer covering has been breached by bacteria, many people would never reach adulthood with all, or even the majority, of their own teeth.

In the same way, the sensitivity of the human skin protects you not only from bacteria and loss of body fluids but also from a host of incidental injuries, like being bitten by a wasp or a dog or even leaning on a hot iron. Normally, we need only touch the hot surface very lightly. A fast muscular reaction called a reflex pulls the limb or finger away even before the impulse reaches the brain. This is why you often feel a burn after your finger is already out of danger. Without pain receptors, you might very well remain in contact with the hot surface until you smelled burning flesh; clearly this is much too late. The first report of a case of indifference to pain was made in 1832. Since then fewer than fifty such cases have been diagnosed.

Is a fan or air conditioner on? There are sounds in the air. Is mother cooking cabbage or baking a cake? You can smell it. Are you chewing gum? You can taste it. The point, of course, is that a great deal is going on around you that you are perfectly content to ignore until some change takes place—as, for example, when the air conditioner goes *off*. In all cases, this is a most useful way of ensuring that the senses will provide useful information on important activities, such as the intrusion of a predator or the advent of food—all the senses are continually feeding information to your brain, which must choose among their messages and pay attention to the important ones.

In the smaller creatures, which have not the space for large brains, the sense organs must compensate for this lack. A hungry frog can be placed in a cage containing flies, a favorite food. If they are alive and moving, he'll catch and eat them sooner or later. If they are dead, or even drugged, and unmoving, he may very well starve to death in the midst of plenty. This is true of many creatures, including that master predator the praying mantis. If he spies a moving spider, his head will turn in that direction. If the spider elects to stand still, as spiders often do, the mantis will remain poised but immobile, apparently unable to "see" his prey. After a while the mantis may advance slowly in the direction he has turned his head, forelegs raised and ready to strike the instant he detects some motion. We see now why "playing 'possum" works in some instances. The preying animal may not be able to see or simply may become uninterested in a motionless prey. In general, of course, being sensitive only to moving objects is a perfectly reasonable approach and makes the best of limited resources. Under natural conditions both predator and prey are in motion. The frog dis-

Praying mantis with prey.

tinguishes between them in a very simple way. If the moving object is smaller, he attacks. If larger, he flees. If the same size, well, anything can happen.

Ducks, wild geese, and vultures have very sensitive eyes. Indeed, they seem to have telescopic vision. Although they have very good hearing, ducks and wild geese especially depend largely on their eyes for protection. The white swan, one of the most conspicuous objects in nature, depends almost entirely on his eyes. He keeps well away from any place that could hide a beast of prey and nests in flat areas where an enemy would be visible for a considerable distance.

Although, as we have seen in the case of the spider, the same creature can be, and often is, both predator and prey, the placement of the eyes can tell us much about his usual role in life. With very few exceptions, the eyes of preyed-upon creatures are at the sides of the head, thus giving them the widest possible view of the world around them. Rabbits, deer, and horses are vegetarians. Clearly they are not predators. All have eyes placed at the sides of their heads. This means that two quite separate images are delivered to their brains.

The alligator and the hippopotamus have eyes set in a raised position on top of their heads. With this periscopic arrangement they can float in the water, almost entirely submerged, and still keep an eye on their surroundings. The woodcock's eyes are placed so far back on its head that he can actually see better to the rear than to the front. This odd arrangement enables him to maintain a lookout while his long bill probes deep in the earth in search of food, mainly earthworms. On the other hand, the bittern's eyes are placed unusually low in his head.

Hippopotamus eyes—an example of "periscopic" vision.

Thus, with his bill pointing skyward to help him blend with the cattails in which he lives, the bittern can still see around him when danger threatens.

By contrast, the eyes of carnivores are almost invariably placed squarely in front of the head and facing directly forward. The tiger, for example, is a powerful and dangerous fighter and has little need for a 360-degree "lookout." The advantage of having two overlapping, almost identical images is that the distance to objects can be estimated more accurately when the brain compares the slightly different images. This is called stereoscopic vision and is useful to predatory animals in that it enables them to judge quite accurately the distance to their prey. In other words, it gives them excellent depth perception. This capability is also useful to tree-living creatures, who

83

Defense Mechanisms

must be able to leap from branch to branch and tree to tree with unerring accuracy.

If you would like to see the advantage of stereoscopic vision, try the following simple experiment. Hold a pencil in each hand and try to get the two erasers to meet by moving them steadily toward each other. Try it with one eye open and then with both eyes open. You will see that it is very difficult to do it in the first case and quite simple in the second.

One creature seems to be able to make the best of both worlds. The chameleon has turretlike eyes that act very much like tail-gun turrets in a bomber. They are located high on his head and can rotate in any direction—independently. While one eye is searching the branches above for his favorite food, insects, the other may be scanning the area below for enemies. When a fly, worm, or beetle is sighted, however, his eyes work together, providing the great advantage of stereoscopic vision. Faster than the

Turret-like eyes of the chameleon can rotate in any direction.

eye can see, the tongue darts out—as much as ten inches for a six-inch lizard—and catches the insect with a sticky secretion at the end. The return is slower and the tongue can be seen to thicken as it comes back with the insect on the end.

Two is the most common number of eyes in the animal world, but certainly is not the maximum. A strange lizard called the tuatara has an extra one on top of its head. The grasshopper has five eyes—two large compound eyes like those normally found on insects, plus three small, simple ones. The spider has eight eyes. And the scallop has a whole pack of them spaced around the outside of its shell.

Because our eyes are so important to us, we tend to think that all animals are as dependent upon these organs as we are. This is not so. Many of the lower forms of life function without eyes or ears, although some may have rudimentary eyes that enable them to move toward or away from light, as the case may be. For example, barnacles are small, shelled animals that live in the sea, many of them firmly attached to rocks, wharf piles, or even the bodies or shells of other marine creatures. The young of the barnacle are harbored for a while in the protective shell of the mother. When released, they seek the light, which of course is upward or toward the surface of the water. This is true of countless other marine types such as worm larvae, snails, and clams. Once near the surface, they are scattered by current and waves. After a while they sink toward the bottom and firmly attach themselves to some object to begin the cycle anew. In this way the offspring of one mother are prevented from settling in the immediate area, which would cause them to be in competition with one another for food. The adult barnacle continues to respond to the light, however, retreating into its shell as soon as the light dims.

As with the eyes, sense organs in general are placed where they will do the most good. Many butterflies and flies have extraordinarily sensitive taste receptors in the soles of their front feet. They frequently eat overripe fruit that has fallen from trees—food that is large enough for them to walk around on. The antennae of the cockroach can be twice as long as his body. Sense organs are advantageously located on them.

At the other end of his body the cockroach has, in addition, two similar but shorter structures called *cerci* (from the Greek word for tails). The cerci are highly sensitive structures covered with tiny hairs; even a light puff of air will send the roach scurrying. Thus, the roach, which has existed essentially unchanged for hundreds of millions of years, is sensitive to activities both fore and aft.

Cockroach. Note long antennae and cerci.

The Senses

Among higher animals, sound or smell may be the most important sense. In such cases the degree of sensitivity is beyond our conception. The kangaroo rat lives in an underground burrow, a maze of tunnels and chambers. This alert little creature listens for trouble in two ways. He can hear through the air in the normal way. But large resonance chambers that take up half his skull also allow him to detect the arrival of visitors by means of vibrations in the earth.

Among the most sensitive "smellers" are wild sheep, deer, and bears. All members of the deer family, for example, depend much more upon their noses for protection than upon either eyes or ears.

Animals, both prey and predator, seem to understand very well that odors are carried by the wind. The tiger, for example, tries to maneuver himself into a position upwind of his prey—for no matter how well he is hidden, his scent will give him away to a deer or an antelope if a breeze is blowing toward that animal.

The hunted creature, of course, tries to do the reverse. The naturalist Roy Chapman Andrews tells of trying to stalk a herd of giant bighorn sheep, who would sleep away the warm early afternoons in plain sight high up on a narrow ridge connecting two peaks. An old, experienced female of the group acted as sentinel, gazing in all directions for a long while. But eventually even she would bed down.

Then Andrews and an assistant would move stealthily toward the animals, making sure that the wind was in the right direction. Yet they could never approach close enough for a shot. Somehow the herd was always up and away in plenty of time.

Defense Mechanisms

Finally Andrews discovered the reason the herd always selected this spot day after day and didn't seem to worry about concealment. It seems the air eddied in a peculiar way and brought odors to the sheep's noses no matter which way the wind was blowing.

But the champion smeller is the moth, especially during the breeding season. With its elaborately feathered antennae, it can catch the few molecules of scent that might arrive from a female two or three *miles* away.

Notwithstanding this fantastic apparatus that has developed in the moth, a quite different kind of sense has evolved to help him in his defense against an unusual predator, the bat; for the bat can dispense with the familiar senses—sight, taste, touch, smell, and sound (as we know it)—and still catch his meal. It is fairly well known that he navigates with a very effective kind of radar, or more specifically sonar (*so*und *n*avigation *a*nd *r*anging). That is, he emits short bursts of high-pitched sounds (ultrasonics). These are reflected from various objects in characteristic ways and are reflected to his big ears. When a bat is airborne, or even preparing for takeoff, he emits rapid bursts of ultrasonics at so high a frequency —40,000 to 80,000 cycles per second—as to be completely inaudible to human ears, which are sensitive only up to a maximum of about 15,000 to 20,000 cycles per second. The wavelength of the bat's frequency range is just a few millimeters, so that the waves are reflected by even very small objects. Thus, a bat can determine the exact location of a tiny bug held in front of him; he can locate every crack that will offer a foothold in a wall; he can even distinguish velvet from paper—all in pitch blackness. In short, the bat can accomplish with his specially adapted

Ears of the bat are adapted for ultrasonic waves.

ears practically everything we can do with our eyes.

Clearly, the moth is faced with a dangerous adversary. For defense he has auditory organs that are sensitive to this same range of ultrasonics. (His cousin, the butterfly, which is not hunted by the bat, does not have any sound sense at all.) If a moth is airborne when he detects the bat's high-pitched sound, he may react in one of several ways. He either will make an abrupt about-face or, more likely, will simply fold his wings and plummet to earth, hoping for refuge there. If preparing for a takeoff, he simply delays his flight and waits till the way is clear.

Some snakes (the pit vipers and many boas and pythons), as well as certain other creatures, carry this emancipation from the usual five senses even further. Even when completely deprived of sight, taste, touch, and smell (snakes do not have hearing), these snakes are able to strike unerringly at a foe or meal.

The word "pit" in "pit viper" refers to two small cavities that lie on either side of the reptile's head, between nostril and eye. These pits are temperature-sensitive organs. But whereas man has about three heat-sensitive nerve cells in a square centimeter of skin, each pit contains no less than 150,000 such cells. The pits are shaped like headlight reflectors and so receive sensation from a limited, cone-shaped area in front of the snake. In other words, the snake can "see" heat waves (infrared) coming out of any creature that is warmer than its background, as virtually all living things are.

This "heat eye" is even useful to the snake during the day to ferret out creatures that are so well camouflaged as not to be visible to the eye. But even more important, by swinging his head from side to side, he can discover not only the presence of an animal but even its size and shape —all in absolute darkness. Thus he may in pitch darkness be able to distinguish a fine-tasting rat from a snake-eating mongoose. An important distinction indeed.

8. Adaptation and How It Happens

"THE soles of our boots wear thin, but the soles of our feet grow thick, the more we walk on them." So wrote the remarkable student of life, D'Arcy W. Thompson in *On Growth and Form*. He was referring to one of the few undisputed differences between living and nonliving things. A piece of granite today is no different from a piece of granite a million or even a billion years ago. But living things tend to develop, to change, in such a way that their chance of surviving any threat or danger is improved.

The various changes, as we mentioned earlier, are called adaptations. We have discussed many in the course of this book. Our interest now is in how these have come about. Although this is not the place for a detailed discussion of evolution, it is important that we understand two terms, *natural selection* and *mutation*.

The first, *natural selection*, is best illustrated by means of an example. Over the past 100 years the trees and buildings in many areas of highly industrial England have become covered with soot and grime. Normal light-colored members of the peppered moth (*Biston betularia*) are well hidden on the bark of clean trees, but stand out like spotlights on the darkened bark of soot-covered trees. In other words, their once-excellent protective coloration is no longer effective in the changed environment. Birds find them more easily and so there are fewer and fewer of them left to pass on the light color to their offspring. The birds may be said to be acting as "selectors."

Does this mean that the peppered moth will eventually be wiped out or at least reduced to a small proportion of the original poulation in these areas? The answer is yes, unless the species can somehow switch to a darker coat.

Well, just as dwarfs are sometimes born to normal-sized human parents, so too are dark-colored moths sometimes born to light-colored parents. These variations from the norm (and particularly the changes in the genetic material that cause them) are called *mutations*. Clearly, the resulting moths are more likely to live and pass on the dark-coat characteristic to future generations than normal light-coated ones. Natural selection continues to take place and the dark-coat style becomes even more firmly embedded in the moth fashion centers of industrial England.

The darker colored moth in the photo is the mutant *carbonaria* of the original light-colored peppered moth. Whereas dark-colored moths constituted only a small percentage of the population in this area a century ago, they now form over 90 percent.

It is important to realize, however, that the switch to a dark coat did not occur because it was needed. Nor was it a planned or deliberate activity on the part of the moths. Mutations of many kinds are occurring in all living things all the time. (One of the most frequent mutations in nature is the one that causes the disease hemophilia in man;

Dark-colored peppered moth stands out strongly on clean bark (above), but blends well with soot-covered tree. Reverse is true for the light-colored variety.

93

it occurs once in about 50,000 births.) In other words, the dark-coat mutation had been occurring in *Biston* right along. But what at one time was disadvantageous is now advantageous, and *carbonaria* is now the prevailing form.

Sometimes the pressures of a hostile environment are so great that there is not enough time for the necessary changes to take place through the natural course of selection. The result may be the disappearance of the population from the face of the earth. Indeed, this has happened often enough that the number of living species is less than one-tenth the total number of those which have lived at one time or another, but which no longer exist as such.

Summarizing, then, we see that living things reproduce and copy themselves from generation to generation. But we also see that there is continuous variation from generation to generation. It would seem that two such factors could not exist side by side, but they do. The first ensures sufficient stability to protect a species from utter chaos, which would arise if each offspring had totally different characteristics from its parents. The second factor provides just enough variability from generation to generation to provide for evolution and adaptation of life forms. It has also made possible the marvelous and fantastic diversity of plants and animals we know today.

Again, it is the combination of mutation and natural selection that has done the job. In the warm, moist, lush times of the late Paleozoic Era the now-lowly amphibian was the ruler. With the arrival of drier times the moist-skinned amphibian was put at a considerable disadvantage. For one thing, there was simply less water around. This could be overcome, and was, with the development of dry skins which permitted the creature to live out of water permanently. Thus lizards and other land creatures began to take over from the amphibian.

But there was another, more serious problem. In cooler weather the cold-blooded animal becomes sluggish. Both his sensory and his muscular apparatus operate at a much slower rate. For example, at a certain point a drop of four degrees is enough to convert a fly from a swift flier to a slow walker.

What was needed was a method by which activity could be maintained even during cold weather. Clearly, when two equal-size and equally well-armed creatures are pitted against one another, the one that is faster has the edge. Out of this environmental challenge arose one of the most important developments in evolutionary history: warm-blooded creatures. Thus, even in the heyday of the giant dinosaurs, there appeared the so-called archaic mammals. Although small and unimpressive at the time —none was larger than a small dog—they were the first of the warm-blooded animals.

These small mammals were the prey of larger reptiles, against whom they could not compete in direct combat during those warm, lush days. Because they were descended from reptiles, they were egg layers, and it was particularly difficult for them to nest and preserve their eggs, which the larger reptiles considered a delicacy. One might even say that the period was one grand and glorious egg hunt. The mammals had a tough time saving their helpless young as well as their eggs, but they obviously were successful in many cases.

One useful adaptation was retention of the egg in the body so that the young were born more highly developed. A second adaptation was a protective association with their young after birth. The birds, who also developed out of the early reptiles, were equally hard-pressed. For some unknown reason they developed the second adaptation but not the first. As you may know, two species of egg-laying mammals still exist, the best known of which is the duck-bill platypus.

In any case, it seems clear that speed and mobility won out over size, for the warm-blooded animals now dominate the Earth and form the group to which man himself belongs. But warm-bloodedness brings with it its own set of problems. For one thing, a narrow range of temperature must be maintained or the animal may freeze to

Duck-bill platypus.

96

death or may die of heat prostration. In a warm-blooded animal, food must supply the engine with enough fuel to keep warm even when the outside temperature begins to drop. The cold-blooded creature does not have this requirement. Pound for pound, then, the mammal requires considerably more food than does the reptile.

All creatures living in colder climates must have some defense against the cold. This can take various forms, of which we shall mention only a few here. Most birds depend mainly on their feathers, and mammals, on their fur. Mammals living in the colder areas almost invariably have thicker fur than those in tropical climates. Experiments have shown that their insulating power is many times greater. Smaller animals obviously cannot carry about such thick fur and are more likely to seek shelter under the snow in the winter.

The common honeybee (*Apis mellifera*) takes an entirely different approach. Although not equipped individually to withstand cold weather, these bees are often found in colder regions. They "huddle" together, those in the center doing a kind of dance to keep warm and generate heat. All change places periodically, thus giving the outside bees a chance to warm up. The queen bee, however, always remains safely in the center.

Many animals, both warm- and cold-blooded, spend part of the winter in a more or less dormant state. Frogs and some fishes bury themselves in pond bottoms, below the frost line; bats and bears retire to caves to *hibernate*. This provides a kind of protection from cold when a normal body temperature cannot be maintained and food becomes scarce. In summer and fall the animals store up enough food in their bodies to tide them over the winter

Bees huddling for warmth.

months, when much of nature goes to sleep, or until food becomes more readily available again. In areas where conditions are particularly severe, the animals may even have in their bodies substances like glycerol, which act in the same way as antifreeze does in a car radiator.

Interestingly, arctic mammals do *not* hibernate, probably because the short, cold summer does not allow sufficient accumulation of the necessary food reserves. Also, the thick accumulation of snow provides some protection against the cold. And finally, there is probably also the danger that the temperature of the creature would fall so low as to actually harm or even kill him.

An analogous mechanism that is useful in the summer or during extremely dry periods is called *estivation*. The African lungfish, for example, breathes by means of lungs as well as by rather simple gills and dies if prevented from reaching the surface to breathe. When it inhabits water that dries out in summer, this strange fish forms a cocoon of mud lined with hardened mucous and lies dormant within it until water is again available.

In extreme cases, creatures such as the vinegar eel may lie without a hint of life for several years. But with the arrival of the proper conditions, they revive and multiply.

As in hibernation, the body processes do not stop but merely slow down. The lungfish breathes during estivation through a little funnel left in the cocoon. Some animals, too, are able to get through particularly difficult times by means of this little trick (without, of course, the mucous-and-mud shell). One such creature, the mohave ground squirrel, remains below ground for more than half the year, for he has periods of intermittent dormancy lasting from late summer to early spring. Thus, his dormancy

99

Defense Mechanisms

involves what would normally be considered both hibernation and estivation. In this case the two processes appear to be aspects of the same physiological phenomenon.

Because of this extreme use of dormancy, the mohave ground squirrel breaks a rule that we stated in the first chapter: he exists in the same area, and in competition with, a similar creature, the antelope ground squirrel. The latter creature is found in many areas where the mohave squirrel is not and is apparently much the stronger competitor for whatever food and drink is available.

Yet because the mohave squirrel is able to stay out of the other's way during times of want, he has, so far, managed to live in spite of the strong competition. Whether this can last remains to be seen.

Hibernating chipmunk.

9. Some Specialists Pay a Price

As long as a defense mechanism, or any adaptation, remains only a part of the animal's life processes and activities, all is well. But if it begins to take over his life or to become extreme in one way or another, chances are he is, or will sooner or later be, in trouble. One of the most interesting examples is the great Irish "elk" (misnamed, for he was actually a giant deer).

The Irish "elk" roamed the countryside about 20,000 years ago during the last "ice age." Arctic conditions and a tundra, or treeless vegetation, favored the development of this magnificent creature. Apart from his great size (he stood six feet high at the shoulders), his most distinguishing characteristic was his enormous antlers, which weighed up to eighty pounds and had a span of more than *eleven feet*. (Skeletons in good condition are on display in the National Museum of Ireland.)

But arctic conditions eventually gave way to warmth and dampness. Trees spread across the lowlands and up the lower slopes of the mountains. The "elk" became extinct at about the same time.

One explanation for his disappearance is that the male may not have been able to penetrate the woods because of the great size of his antlers. Others claim that this is not

Irish "elk" reconstructed in painting by Charles R. Knight.

correct because the "forests" of the time were rarely more than three or four feet high. They claim that the real reason was the development of treacherous marshland, which ensnared and engulfed frequent victims. For the great weight of the antlers was carried on the front hoofs, which remained small and unable to support the great load on the soft ground. The fact that more than a hundred "elk" heads were found close together at Ballybetagh Bog in County Dublin lends support to the latter idea.

In either case the magnificent antlers, which served in good stead in one set of conditions appear to have brought about the extinction of the creature in another.

Of all adaptations and specializations, large size is clearly the most impressive. Just as the predator has an advantage in being fast and agile—qualities that go with relatively small size—the prey has an advantage in being big and powerful. The large animal is better able to fight off his attackers or, even better, to intimidate them by his very size. We have already mentioned that a large creature may not be harmed by a venom that would paralyze or kill a smaller one. The large animal—the rhino, for example—can also carry a thick hide, which helps him cope with insects and other dangers that large size alone cannot help.

Yet nature's record tells us that size alone not only is not a great help but is often a danger or at least a decided disadvantage. The most spectacular case is that of the big dinosaurs. They were once rulers of the earth, but there is nothing left of them but a few small relatives. Lest you think that their major problem was the fact that they were reptiles and hence cold-blooded, we need only mention the *Baluchitherium*. Although a far cry from the sixty-plus tons of the *Brontosaurus*, this rhinoceros-like mammal of 30 million years ago was quite an impressive beast nevertheless, for he weighed some ten tons, stood about eighteen feet at the shoulder, and could reach leaves with his mouth that were twenty-five feet above the ground. Yet he too has disappeared from the face of the earth, along with the mammoths of the ice age, the dinosaurs, and a number of other giants in nature.

No single reason can be given for the extinction of so many large creatures over the ages. Perhaps with change in climate the large vegetarians simply ran out of food, which would have had to be ingested in gigantic amounts to feed those huge bodies. The larger carnivores, in a similar manner, would have needed to catch huge amounts of smaller creatures to make up for the loss of larger ones. In any case, it is clear that extreme size alone is not a sure-fire defense mechanism and may very well lead to trouble.

There are, of course, many other kinds of specialization. Consider, for example, the koala, a funny little climbing mammal found in eastern Australia. His diet is two or three pounds of leaves a day—but *only* from the eucalyptus tree. What happens if a blight kills off all the eucalyptus trees? The koala is quite likely to starve and die out even if surrounded by many other kinds of leaves.

Thus, in terms of prediction, the most likely candidates for a long future and a diversity of descendants are those creatures that are now relatively small and unspecialized. Among mammals the rat is in a particularly strong position. He is omnivorous (eats both meat and vegetables, indeed almost anything); he is small and can exist on relatively little food; he has a higher productive rate; and, finally, he has a reasonable development of the various mammalian characteristics, including a quite adequate brain. He has also achieved a kind of balance with other natural forms.

I have already mentioned two of the oldest animal species that have survived in essentially the same form over the ages, namely, the scorpion and the cockroach. Both are small.

Few of us would miss these should they be wiped out by some miracle. Unfortunately, man has been more successful in killing off a number of far less offensive members of the animal world. Among the most poignant examples are those of birds who grew too large and lost their capacity for flight. I mentioned earlier that this happened mainly in areas where natural predators were absent. As has happened so often, however, man appeared on the scene and upset the natural conditions. The last two great auks were cut down off the coast of Iceland in 1844; the dodo (who appears in Lewis Carroll's *Alice's Adventures in Wonderland*) has been extinct since the late seventeenth century, killed off on Mauritius, an island

Razorbill auk.

Passenger pigeon.

in the Indian Ocean; the great moas of New Zealand, the largest of which attained a height of thirteen feet, were killed off 400 or 500 years ago; and the elephant birds of Madagascar, which may have weighed as much as 1,000 pounds, were hunted to extinction even earlier, some 600 or 700 years ago.

But man's viciousness is not only vented on large and flightless creatures. A surprising example is that of the passenger pigeon, the last of which died in a Cincinnati zoo in 1914. What is surprising is not only that these birds were good fliers but that they once existed in such fantastic numbers that the great American naturalist-painter J. J. Audubon once wrote of a day when "... The air was literally filled with pigeons; the light of noonday was obscured as by an eclipse . . ."

But if the pigeons existed in such numbers, could they actually be wiped out by man? The answer, strangely enough, is yes. For they were killed in flight, while resting, nesting, or whatever. Audubon told of a man who caught and killed upward of 500 dozen in a single day, using a net.

The point here is that even though the birds did not trade their specialization, flight, for the opposing one, large size, they nevertheless were not able to resist the onslaught of the most merciless (and best-armed) killer who ever lived: man. The rat and the cockroach, which are both smaller and "generalists" of the highest order, are still very much in evidence in spite of man's continuing campaigns against them.

It is worth pointing out that not all of the members of a species have to be killed for it to be wiped out. Should the numbers of any species be reduced below a certain figure —which may be in the thousands or even millions—the species is marked for total destruction. For remember that every animal has its natural enemies to contend with: disease and hunger, as well as predators. If man reduces the animals to this figure, which is called the threshold, these natural enemies take care of the rest. This is what happened to the passenger pigeon.

A somewhat happier outcome can be reported for the equally harried but smarter and more adaptable sea otter. Various species of otters (relatives of the weasel and skunk) enter the sea at times, but only one species, the sea otter, is entirely marine. He is a large animal, four feet long and weighing up to eighty-five pounds, and receives

107

Defense Mechanisms

rigorous protection from both the Russian and the United States governments. He is rare today, having been hunted ruthlessly in the past for his thick, dark-brown fur. In 1916, there were so few left that a law was passed forbidding the killing of sea otters. But the clever creatures had already taken steps to protect themselves. Only a hundred years ago these otters lived largely on beaches at the northern edge of the Pacific Ocean. There they found their food, sunned themselves, mated, and gave birth. Although they spent much time in the water, the beaches were "home."

Man made that home unlivable. Had the otter been less adaptable, he might have gone the way of the passenger pigeon. The otters simply moved out and have become sea animals exclusively, although they never move far from

Sea otter, using his stomach as a table, eats a codfish.

shore. They learned how to wrap themselves in the fronds of seaweed to keep themselves from being swept away from the rest of the herd while they sleep. They learned to find all their food in the water. The female even gives birth to her single cub in the sea and hugs the young one to her chest afterward.

Their food consists of crabs, fish, mollusks, sea urchins, and seaweed. Whenever the sea otter finds a shellfish, he also picks up a flat rock and then swims to the surface. There he lies on his back, places the rock on his chest, and cracks the shell on the rock! It may take two or three times, but eventually the shell gives way and the meat is exposed.

Along with the Rocky Mountain bighorn and various other creatures, the sea otter has managed to exist by moving into another ecological niche where the competition is less fierce. Fortunately, he is clever and adaptable and apparently has not given up a great deal for his move.

Some animals, on the other hand, have had to pay a high price indeed in adjusting to a new environment. Among the strangest are those found in various damp, silent caves around the world. Within such caves, formed by the action of water, are rivers, wells, or sinkholes where water collects. Although the caves may be perfectly dark, we can find, if we are extremely quiet, a number of creatures inhabiting the area. Most common are amphibians, fish, crayfish, and other crustaceans, as well as many kinds of insects and spiders.

Because the caves are dark, most of the cave animals have lost their protective pigmentation, which is no longer needed. Thus, some are white, others even transparent. And they have given up something else that is no

Mexican blind cavefish.

longer needed: their vision. Some are simply blind, but others have lost their eyes altogether. A number of the creatures have apparently retained some sensitivity to light, which they shun. Many even die under prolonged exposure to light. The blindfish of Mammoth Cave in Kentucky, for example, is both colorless and blind. This fish seeks the dark and is greatly disturbed by direct sunlight, a flashlight beam, or even a lighted match. These cave dwellers move around like white ghosts in the silent blackness of the cave, and only the most careful and perceptive humans are likely to see them.

In compensation for their loss of vision, they have developed some of their other senses to a high degree. The colorless cricket sports preposterously long antennae. The blindfish possesses extremely sensitive nerve endings that alert him to the various types of movement in the surrounding water. Should an insect or even a smaller fish of the same species happen by, the blindfish swims unerringly to the prey and eats it. Because no vegetation exists except a few scattered molds and fungi, all the animals are carnivorous and prey on each other as well as on other small creatures that are blown or wander into the cave by accident. Here the highly developed nonvisual senses of the cave dwellers give them the advantage.

The blind crayfish is a common cave animal. He differs from his larger out-of-doors cousin the lobster in being blind, deaf, colorless, and slender-bodied. However, the blind crayfish has highly developed senses of feeling and smell. He spends most of his time in the water and can also detect the slightest movement in the water. He will flee at the slightest ripple or even a spoken word. Better a hundred flights too many than one too few.

This creature is typical of cave dwellers in his ability to cope with lean diets and even semistarvation. All cave animals are small and have small energy demands. There is little unnecessary movement or noise. On the other hand, adjustment to the constant conditions of the cave has made them more delicate and sensitive to change than their outer-world counterparts. The typical cave animal cannot withstand large fluctuations in temperature or force of their pool or river waters. Should large changes occur, disaster results.

10. Defensive Behavior

ONE day a shepherd was driving several hundred sheep through a gate. A boy sitting nearby put out his foot in front of the leader, who promptly jumped over it. The next sheep did the same and so did the next few in line. Then the shepherd asked the boy to withdraw his foot, which he did. Nevertheless, the following sheep jumped when they came to the same spot, even though there was nothing to jump over.

This following instinct of sheep has been pointed to as an example of the foolishness of the animal, for the flock will do whatever the lead sheep does. If he turns a sharp corner, they will too. If he takes a long leap, they will too and will land almost exactly where he landed. It reminds one of the youngsters' game of "follow the leader." No matter how silly the behavior of the leader, the others in the game must do the same.

But look at it from the sheep's point of view. It is quite unlikely that the lead sheep is playing a game. If he does something, there is probably a good reason for it. With certain exceptions, animals have neither the time nor the energy for games; they are too busy obtaining food, defending themselves, finding shelter, and so on.

The second point is that domestic sheep are descendants of mountain and hill creatures. These were often pursued by wolves, and their flight often took them along paths they had never seen before. These paths may well have been narrow rocky ledges, with room for only one sheep at a time. The leader comes to a ravine twenty feet wide. He leaps, makes the other side, and races on. The next sheep could never have seen the ravine in time to prepare himself for the jump; for the view was blocked by the leader, whom he is following close behind. Nevertheless, he leaps without hesitation and lands in the same spot as the leader did. So in quick succession all the sheep are able to cross the gap. A wolf following close behind cannot make such a leap; nor, because of his momentum, can he stop himself. Therefore, he may very well plunge to his death.

So although it seems silly for a whole flock of sheep to jump when unnecessary, we see that it may be a carry-over from an apparently inborn behavioral pattern that proved a lifesaver many times in the past. Although modern sheep have been domesticated for 2,000 or 3,000 years, their peculiar defense mechanism has not yet been lost.

Other types of following response are seen in a number of different creatures. For example, the gnu, a relative of the antelope, is a favorite food of the hyena. Because adult gnus can outrun hyenas, the latter will normally look for newborn or orphaned calves. Having spotted one, he is generally able to capture it in a quarter to half a mile. Older calves are more difficult to capture and may last for several miles.

The favorite prey of the hyena, however, is a newborn orphan, for it will walk right up to the hyena—or any-

113

thing else that moves! Under normal conditions this following response is useful, because it is almost always the mother who is seen first. If the mother is killed or frightened off within the first few days, however, the gnu is obviously in serious trouble, even though he is able to stand and run within seven minutes after birth.

For the first few days, the gnu cannot keep up with his mother at top speed. However, a hyena is not likely to be able to capture one when mother gnu is around, for at the first alarm the mother immediately leads the calf into the nearest herd, against which the hyena is no match.

Many animal species utilize this kind of cooperation as their main means of defense. Among the best known and best organized are the baboons, which are found in groups ranging from single families to a hundred or more. Whatever the size of the group, a very rigid social organization is in force, dominated by from one to three large males. When on the move, the lesser males and some large youngsters go first, followed by the females and the other older juveniles. In the center can be found the females with smaller offspring and the dominant males. The rear group is a repetition of the forward one. When the dominant males stop, the whole troop stops. When they move, the troop moves. Even while resting, the troop maintains at least roughly this defensive formation.

The dominant males are the big guns. Any enemy foolish enough to attack the troop will sooner or later have to deal with them.

The leopard is the traditional enemy of the baboon. Even when alone a large baboon can give a good account of himself with his enormous teeth, which can reach a length of two inches. Normally the leopard is the stronger

Baboons maintain a defensive grouping even when on the move.

and wins. But when in a band, the baboons are not likely to be in serious danger from even the most powerful leopard.

Although baboons usually pay little attention to other animals, they sometimes form a sort of loose partnership with herds of impala. The baboon's keen eye and the impala's sharp nose make an excellent combination. Each knows the other's warning signs and flees when the signal is given.

In another kind of mutual-benefit arrangement, oxpeckers perch on the backs of elephants, rhinoceroses, and other African mammals feeding upon ticks and lice. The mammals derive benefit from the association because they

115

are warned of the approach of danger when the birds fly away.

The general term for association, in which two different living species live in some kind of association with one another, and when both benefit, is *symbiosis*. An unusual example of symbiosis is seen in the relationship between the sea anemone and the clownfish. The latter is slow and highly conspicuous and has no obvious means of defense. The sea anemone, on the other hand, is a fearsome creature whose waving tentacles can deliver lethal stings to almost anything near its size. Yet the sea anemone does

Sea anemone and clownfish—an example of symbiosis.

not seem to mind the presence of the clownfish at all. As a matter of fact, observation indicates that the association benefits the anemone too, for it seems to enjoy better health when there are clownfish around. One possibility is that the clownfish, when chased by predators, leads them to their death among the tentacles, thus helping to provide food for the anemone.

The anemone is a kind of home base for the clownfish, from which he makes his forays for food. The area over which an animal or group of animals wanders is called a range. Often he shares the range with other creatures of his own or different species. Many animals, however, stake out a *territory* or home range from which they try to prevent all other creatures from entering. Wolves, for example, defend the immediate area around their den, although they have a much wider hunting range. The male wolf marks off the boundaries of his territory in a way that every owner of a male dog will recognize (by urinating on certain markers such as rocks, tree trunks, etc.). This restriction refers, of course, to creatures near the wolf in size. He does not, for example, waste energy in trying to keep out insects.

A well-developed system of territoriality is useful to the species as a whole because it tends to limit the population of the species in any one area. The animals, therefore, spread out and maximum use is made of their possible range. Competition for limited resources is also diminished in this way.

Territoriality is related to another behavioral characteristic of interest to us. Each animal has a *flight distance*, the distance to which the animal will allow potential enemies to approach before he runs away or takes other defensive action such as freezing into immobility, shamming death, diving under the water, and so on. The antelope is a skittery creature who will run if he detects the presence of a man as much as 500 yards away—the length of five football fields. This distance may vary somewhat depending on such factors as terrain, vegetation, and the physical state of the animal or herd. Smaller animals have smaller flight distances; a sparrow will normally allow a man to come as close as six feet.

It has also been established that many animals have in addition a second, smaller territorial zone. This is a roughly circular area with the animal in the center and a radius called the *critical distance*. The intrusion of an enemy into the critical zone may turn flight into attack. The animal fights "like a cornered rat." Flight distances and critical zones must be carefully considered by zookeepers in order to keep peace among wild animals in a zoo.

Animal trainers in circuses must also keep these distances in mind; indeed, their acts in many cases depend on them. Normally a lion in a zoo will run from an approaching man until he meets a barrier. If the man continues to approach, the cornered lion will reverse direction and begin to stalk the intruder. This stalking is so deliberate that the lion will climb right over an intervening obstacle in his path. The obstacle might be a stool in a circus animal act. To get the lion to remain on the stool, the lion tamer quickly moves out of the critical zone. The lion simply stops pursuing (assuming he is not hungry) and remains on the stool. The elaborate protective devices —chair, whip, and gun—are generally used for effect only. Even though the lion is quite high on an evolutionary scale and is quite an intelligent animal, a lot of his behavior remains instinctive and is thus invariable under certain conditions set up by the trainer. After a while, of course, the lion may begin to do things because he knows he will be rewarded for them. Nevertheless, the trainer keeps the lion's natural behavioral patterns very much in mind.

Animal instinct shows up in a number of interesting ways. A large kangaroo and a stag were once placed in the same enclosure in a zoo. As long as the kangaroo sat on all fours, which he sometimes does, all went well. But whenever the kangaroo rose to his typical two-legged stance the stag attacked it.

This is interesting from several points of view. First, although the antlers of a stag may be excellent weapons, they are not often used as such, even against beasts of prey. It would be like a weapons collector soiling his best sword on a troublesome weasel when he has a gun or a trap to take care of the intruder. For in general the antlers of a stag are a sexual symbol, like the fancy feathers of a peacock. When cornered, stags are more likely to rear up like a horse and use their hoofs for defense. Apparently, then, the stag in the enclosure took the two-legged stance of the kangaroo as a threatening gesture; the response was inevitable.

This does not mean that the stag does not use his antlers for fighting. He does, but usually only when battling another stag of his own species (called *intraspecific fighting*) and only in a very special and ritualized way. Wild goats, mountain sheep, and antelopes fight in a similar manner—that is, using foreheads and horns, and with each species using its horns in a very specific way. They

119

may crack heads, engage in a pushing match, wrestle with locked horns, and so on. But almost never will they try to gore their opponent, regardless of how sharp their horns are—and in such creatures as the oryx antelope, the horns can be very sharp indeed.

As a matter of fact, there is some correlation between the effectiveness of an animal's weapon and his reluctance to use it, particularly on members of his own species. There is a very good reason for this. If armed animals did use their potent weapons on their own kind—for establishing territory, winning mates, or setting up a pecking order (determining who is boss)—they might very well wipe themselves out in short order.

It should be stated, however, that starvation conditions may overcome this inhibition. After all, all behavior—even in the simplest creatures—represents an attempt to adjust to some sort of change or stimulus. Conflict is gen-

Male black bucks engaging in intraspecific fighting.

erally produced by an unpleasant or unstable environment. Examples are the existence of pain or lack of food. A pair of mice that coexist very nicely under favorable conditions become extremely competitive if offered only a single pellet of food when hungry.

In such case the stronger or the dominant mouse will get the food. We may be repelled by the idea that "might makes right," but under natural conditions it works to the benefit of the species. Instead of all the animals starving under poor conditions, a few (the strongest and most aggressive) will survive! These can then pass on their genetic characteristics to their offspring.

Under normal conditions threatening gestures, feinting, prolonged glaring, and ceremonial displays often replace actual combat. The songs of many birds are not, as is commonly thought, a sign of pleasure; they are almost entirely a replacement for conflict in claiming territories. In howling monkeys a special voice box has evolved that can rival a strong siren; and they use it—in groups—for the same purpose. (Howlers are probably the loudest-voiced creatures in existence.) The peacock shows his dominance over other peacocks by displaying his magnificent plumage.

As a matter of fact, the bright colors in birds are almost always found among the males of the species. The females are generally quite drab by comparison. This helps to make the female less conspicuous, which is particularly important when incubating her eggs. It is interesting, however, that the male phalarope is an exception. He is less highly colored than the female—and is one of the rare species where the male rather than the female incubates the eggs.

121

Defense Mechanisms

As always, the most important consideration is the survival of the species as a whole. The inhibitions against one animal killing another of his own species naturally serve this purpose very well. For this to work, however, it must be a reliable characteristic. For example a treacherous person is often called a snake. This is quite unfair to snakes because they can be counted on to adhere to the rules of the game. In spite of the fact that one rattlesnake could kill another with a single bite their battles are invariably limited to a kind of "Indian wrestle" in which one tries to "pin" the other. The loser is, however, given his freedom.

The rules to which the European lizard *Lacerta agilis* adhere are rather curious. The fight begins with a short introductory "display," after which one grasps the neck of the other in his jaws. The second waits for the grip to

Timber rattlesnakes "Indian wrestle."

loosen and then takes his turn at biting. The "fight" continues until one decides he has had enough and runs away. Often, however, it is the biter and not the bitten who gives up and flees. Perhaps he recognizes the superior strength of his adversary by the latter's unyielding resistance while being bitten as well as in the strength of his bite.

In each case of intraspecific fighting, there is a characteristic action that says, "OK. I've had enough. You win." At this point the winner desists and the contest ends. Among iguanas, the struggle ends when one of the lizards assumes the posture of submission, which is that of crouching on his belly. The winner immediately stops charging and waits in a threatening, stiff-legged stance until the defeated lizard retreats.

These forbearing reactions by the winner are not remarkable examples of self-control in the same sense that we would think of them. In the same way that the stag automatically attacked the kangaroo when it stood up, the winning animal in intraspecific fighting *must* stop when the loser assumes the loser's posture. This occurs in various ways. A defeated wolf bares his neck, his most vulnerable spot, for the lethal bite. But it never comes, for that action causes in the victor an overwhelming urge to urinate. This gives the defeated wolf time to get away. Again we must state that there would be no more wolves —or stags or lizards—if there were not reliable inhibitions to prevent these creatures from harming one another. Life in the wild is tough enough as it is.

As always, there are exceptions. Doves, the very symbol of peace, do not, oddly enough, have instinctive gestures of surrender that would inhibit the attack reactions of a stronger individual. The species has survived because in nature escape by simple flight is generally possible. But in captivity a weaker bird may be pecked to death by its rival—one of the very few cases in nature where this will take place. So much for the bird of peace. Whoever chose the dove as the symbol of peace apparently did not know much about animal life. (It is interesting to speculate that the present study of natural history probably goes back to primitive man, who *had* to know animal habits if he was to be able to exist at all.) On the other hand, crows, ravens, and other birds with powerful or sharp beaks are inhibited from pecking at each other's eyes, which would blind them and leave them helpless.

Indeed instinctive behavior is far more highly developed in birds than in most other backboned animals. A male robin defending his territory does so by means of a territorial display. But he will display as vigorously to a small red feather mounted on a wire as he will to another robin. The only thing that is important to the territorial male is the patch of red at a certain height.

An even clearer example of the rigidity of instinctive behavior is seen in an experiment performed on the Sphex wasp, which hunts crickets. When the female Sphex has taken a paralyzed cricket to her burrow, she leaves it on the threshold and goes inside for a moment. Apparently, she is checking to see that all is well. Then she emerges and drags the cricket in.

Some years ago, one experimenter moved the cricket a few inches away while the wasp was inside. The wasp came out, moved the cricket back to the threshold and went inside again—at which point the experimenter moved the cricket away again. The wasp repeated her action. The experimenter did this forty times, always with the same result; the wasp never responded by pulling the cricket straight into the burrow.

The sequence of actions—drag cricket to threshold, pop into burrow, pop out, pull cricket in—is like the count-down for a big rocket. Mess up one step in the sequence, and we must start all over again.

This type of phenomenon occurs because insects do very little thinking. Thus, a particular object may release an elaborate train of behavior, while another is paid no heed to at all. Indeed, we can even compare this to the same automatic way that light of a certain wavelength hitting our eyes gives rise to a reaction that we know as the sensation of red, while light of a slightly longer wavelength is not perceived at all. (Remember how the robin displays to a *red* stimulus?)

The insect is turned out with a battery or repertory of these behavior patterns, just as a music box may have a few different tunes but no others. Even the most complex activity can be broken down into a set of simpler units, each of which is *released* by the previous activity or some other stimulus. Sometimes the behavior of an animal, even though it seems complex, can be accounted for in terms of biochemical changes induced by particular stimuli. How can we account, for example, for the fact that mice tend to spread out—that is, to decrease the population density —when food is scarce? Is this something they all decide to do during some town hall meeting? Perhaps. But another answer is more likely. Experiments have shown that well-fed mice are "lazier" than starved ones. Or, to put it the other way, it has been found that starved mice are far more active than well-fed ones.

125

Defense Mechanisms

One would think at first that it would be best for them to do as little exercise and running around as possible, in order to conserve their energies until things improved. But mice are short-lived creatures. It is rather unlikely that things would improve fast enough to be of use to them. Much better for them to spread out and seek new and more fertile territory. And sure enough, experimenters who kept track of the number of turns of the exercise wheel in mouse cages found that hungry mice "ran" far more than well-fed ones—a built-in mechanism that helps spread the creatures out in time of want.

Clearly this is not reasoned behavior, because they responded in a similar manner in the cages, where it was of no use at all. Apparently, starvation releases in the bloodstream of the mice a chemical substance (a *hormone*) that stimulates them into activity. (In man the thyroid gland is known to be a controlling factor in the level or rate of activity.) In a similar manner, feigning death may be an involuntary nervous response in such creatures as beetles, spiders, and even snakes and birds. To keep things simple, we shall continue to refer to such behavior as instinctive. Similar behavior is seen higher up on the evolutionary ladder too. But here, as for example, in hyenas, the behavior may be a result of cunning.

The difference between instinctive and learned behavior is clearly seen in the defense mechanism of the puff adder or hognose snake *Heterodon contortrix*. The puff adder is short and thick and has markings like a rattler's. If threatened, he widens and flattens his head; he rears up, hisses, and threatens to strike. Yet, although he looks quite fierce, he is really quite harmless. If the first part of his act doesn't scare away his enemy, he goes into convulsions.

He actually writhes in apparent agony. His head, with mouth open and tongue hanging out, twists to one side. After a few more such activities, he rolls over on his back, "playing possum." So far he's put on a very good act. But now we see how rigid his act is. For if you turn him over onto his stomach—he immediately rolls over onto his back again. Apparently, no one ever told him that when he's dead, he's supposed to stay dead. But maybe his enemies are not much smarter than he.

If the predator does leave him alone and wanders off, the snake very soon begins to show more reasonable signs of life. He looks about and if the coast is clear, he slithers away as fast as he can.

The advantage of instinctual behavior is that it provides a set of prefabricated answers to often-encountered stimuli and dangers. This is particularly important for creatures that may be left alone minutes after birth or even, in the case of some egg layers, before birth. Clearly, if the newborn offspring had to learn to duck, or run, or fly, or swim, or play dead, or even discover who his enemies are, the species would be in serious trouble. Thus chickens peck, rats mate, and canaries sing the canary's song without experience. (Among some birds the unlearned song is a basic one that becomes more complex if the bird is reared among his fellows.)

Experiments with birds even seem to indicate that predatory birds are recognized without training. Predatory birds generally have shorter necks than nonpredatory types. In experiments along this line, Dr. Niko Tinbergen of Oxford University used a black cardboard model of a bird, one end of which was short and the other long. Using inexperienced ducklings, goslings, and chicks on a lawn, he passed the model, via a wire, over their heads. When the youngsters saw the model moving with the long end up front, there was no commotion. But when the model passed over with the short end leading, they scattered and hid. It would seem then that the short neck of a flying creature is their automatic cue that there is danger in the air. Here the instinctive behavior is apparently associated with some kind of inherited structure of the nervous system of the creature.

Birds do learn also, of course. The mimicry of the viceroy butterfly would be useless if they didn't. And as we move up higher on the evolutionary scale, learning becomes more and more important. Among dogs and wolves, for example, the young appear to learn the meaning of specific sounds or facial expressions of the parent only gradually; repeated bites may be necessary to impress upon the youngster the fact that not every form of growl is an invitation to play.

This kind of education requires two major characteristics. One is a relatively stable social order. Clearly, as we mentioned earlier, a youngster who is abandoned very early in life would be in trouble if he had no preset answers to certain dangers. Dogs, wolves, and baboons are social animals and the young are cared for and taught.

The second requirement is different but allied. The nervous system, which includes the brain, must develop slowly. If it develops and "hardens" too quickly, the amount the youngster can learn is very limited. Learned behavior is, of course, more flexible and will serve in cases where instinct might cause trouble. But learning implies making mistakes, which in the wild can also be very costly. And it takes time for learning to occur. The young howler monkey will not wander off by himself for fully six months of his young life, and even then he never strays far from his mother. He cannot be considered completely independent until he is three years old. Among lions and other primates, there is also a relatively long childhood, and a great deal of training takes place.

The extreme example, of course, is man—at whom we shall take a closer look in the next chapter.

11. Communication, Culture—and Man

THE skunk's chemical spray, once used, does more than just drive an enemy away. It also serves as a warning signal to other skunks that there is trouble afoot. Indeed, chemical alarm signals are a common type of warning in the animal kingdom. Fish, for instance, give off an alarm material when injured that causes other members of the species (and other species as well) to flee. This material is extremely potent. As little as 0.002 milligram of fish skin (about 0.01 square millimeter) in a 3.5-gallon aquarium can alarm a group of minnows. Not too surprising is the discovery that the alarm material is attractive to sharks.

However, it should be noted that these alarm signals are secondary effects. That is, the fish does not deliberately give up a piece of his skin to warn his fellows. Similarly, the primary purpose of the skunk's spray is self-protection, not alarm.

There are many cases of the use of protective signals where this is not so. The beaver, when frightened or disturbed, often produces a very effective alarm signal that has nothing to do with self-protection. Indeed, it may very well call attention to the beaver who produces it. He gives the signal by hitting the surface of the water with his powerful, flat tail. The result is a resounding slap that echoes through the surrounding area—and the immediate disappearance of every beaver within hearing distance.

The pronghorn antelope depends largely upon his sharp senses and high speed for protection. But he also possesses a patch of long, pure-white hairs on his rump. When the animal is frightened, the patch expands into a bright round disc. Rapid opening and shutting of the disc can send signals like flashes on a ship's heliograph. In flat areas the signal can be seen for miles. Peacefully grazing pronghorns will see this signal from the proverbial corner of their eyes; after throwing up their heads for the briefest instant, they dash away.

Young pronghorn antelope "signaling."

Defense Mechanisms

Among the most complicated alarm systems are those found among the social insects, such as ants and bees. Ants, and some termites as well, appear to send danger signals by three different routes: (1) chemically, by production of certain odors; (2) through the sense of touch, via wild alarm dances; and (3) by production of rasping sounds or tapping of the body against the side of the nest. Many species of ants and termites have washboardlike ridges on the abdomen (the rear section of the three-part body). By rubbing the abdomen against the thorax, or middle part, they can produce high-pitched sounds. In some species, at least, this is done only when the insects are disturbed, and an alarm is spread throughout the colony.

It is tempting to think that the animals and insects mentioned above are warning their friends in the same way that we may yell, "Look out!" when danger threatens a friend of ours. But in general this is not the case at all. The warning signals are more likely to be simply a nervous reaction of some particular kind resulting from the fright of the animal. We might compare it to the sweating, strong heartbeat, or chattering teeth of a frightened human. In a certain sense the animal cannot help but produce the alarm signal, even though it may bring the predator straight to him. The important thing is that this particular reaction has a useful function in that it acts as a warning signal and so is useful to the species. The chances are that it started as something quite inconspicuous at first, i.e., in the early days of the animal's history. Then, because of its survival value, it evolved to its present highly developed form.

Still, one can hardly help calling it a form of communication, for it certainly does communicate information. Some game bird chicks, for example, "freeze" at one call from the mother hen and will not move until the "all-clear" signal is given. Oddly enough, the *absence* of such a signal can also be used as a kind of communication. Some time ago the German biologist Albrecht Faber noted that when he walked about in a field, all the grasshoppers jumped at his approach. Here we have a case of an alarm signal that is similar to that of the impala. The very jumping of the creature is the alarm. Yet, Faber noted, grasshoppers jumped about all the time without alarming the others. How could this be?

Careful investigation revealed that before jumping, each grasshopper gives a short set of clicks, which Faber called the departing song. These tell the others in the area that he is going to jump. The rustle of the jump, when it occurs *without* prior notification, alarms the grasshoppers.

Sounds are particularly well suited for alarm signals for several reasons. One is their intermittent nature. Color and pattern are useful in the reverse role—namely, concealment—because they are always there. Sound, by virtue of the fact that it is intermittent, makes it ideal for alarms. There is, in a sense, a surprise factor. Sound also can go around corners (which light cannot do) and can travel long distances even when there are obstacles in the path. In addition, sound works as well at night as during the day and can operate against the wind as well as with it (as smell cannot). And finally, the great number of possible variations in tone, pitch, loudness, and pattern make it a marvelous tool for warning one's fellows.

These characteristics also undoubtedly explain why the basic method of human communication is by means of sound rather than by visual or tactile (touch) signals. Among some of the higher mammals, however, visual signals are very much used. Indeed, both primates and man utilize facial expressions to a very great degree. And the bared fangs of a dog are well understood by all.

In other words, visual signals are very useful for the communication of basic information. But they are severely limited in flexibility and capacity and could not possibly have provided the raw materials for practical human communication. Anyone who has watched two deaf-mutes communicating with sign language must marvel at their speed and dexterity, but can see immediately how limited it is in comparison with the marvelous flexibility of speech. And, it must be remembered, the communication process of two deaf-mutes depends upon a prior knowledge of *language*. It is awkward because it is a substitute for speech.

Baboons have a wide range of both vocal and visual signals, as have many of the other primates. But these cannot be considered even as a step toward human language. Monkeys and apes, says anthropologist Jane Lancaster, communicate emotions, dominance, submission, warnings, and so on. But even though these vocalizations come closest of all those produced by animals to the kind of communication practiced by man, they are still closer to those of other animal life than they are to our words and sentences. Monkeys cannot, for example, communicate information about their surroundings. They may be able to indicate hunger, but they cannot say, "I'm tired of bananas. Please get me a handful of nuts today." Still less can they say, "I went for a walk by the river yesterday," for all animals live in the present. Even storing up nuts, as certain animals do, is an instinctive operation, a response

"*My man don't wrestle till we hear it talk.*"

to an outside stimulus, not really planning for the future.

The great chasm between the communication abilities of man and ape was nicely pointed up by the cartoon that appeared in *The New Yorker* magazine.

To this wrestler, at least, the only sure way of distinguishing between man and beast was by means of speech. No matter how poorly a human may speak, there can be no question that the sounds are those of a human and not those of even the most intelligent ape.

Further, man is the only creature who can produce and pass on information that will last beyond a single lifetime, for the primary function of language as we know it is the communication of learned material so that each generation does not have to solve problems anew. Man accumu-

135

lates, and passes on to his offspring, a "culture." Here we have arrived at the crux of the matter.

After all, man is slower than the cheetah, weaker than the lion, and smaller than the elephant. He can't climb as well as a monkey, swim as well as a fish, or fly as well as a bird. Yet he is unquestionably the most successful large form of life on earth. (When it comes to creatures such as insects, mice, and rats, the picture is not quite so clear.) No other large mammal even comes close to him in numbers.

How did man manage to come out so clearly on top? It seems likely that one of his first and most successful defensive weapons was his use of fire. By bringing fire to the places where he lived, he created a safety zone for himself that was far superior to anything other creatures could manage, particularly at night. For the first time, he could turn night into day and had a way of keeping the powerful nighttime prowlers at bay.

But for a long time he had to depend upon natural fires for his supply. He could take a burning brand from a forest fire caused by lightning. And he had fire as long as he could keep it burning. But if it went out, he was out of luck until the next natural fire occurred. There are indications of his use of fire for cooking and warmth that go back 750,000 years. But the first evidence of his *making* a fire—a round ball of ironlike material with a deep groove in it produced by repeated striking to create tinder-igniting sparks—dates back only 15,000 years!

This ball was a tool, and an important one at that. Others were more obvious means of defense and of course offense. Once man had the club, dagger, sling, spear, and, later, the bow and arrow, he was well-nigh invincible. But it is widely believed that these developments, as well as his ability to produce fire at will, could not have taken place until the development of at least some primitive form of speech. Even man's intelligence is thought to have developed along with, not prior to, his use of language. For in contrast to animals, humans use language as a tool for thinking as well as a means of communicating.

Most animals that are ground living and solitary ("loners") must be silent and stealthy in order to survive. When our ancestors, the early primates, took to the trees, vocal communication became extremely important. For it was the only way the family band could keep together while traveling through the tall, dense primeval forests. As we have already seen, cooperation and social existence are excellent defense mechanisms. Man, too, is a social animal. But think of how much more effective cooperation becomes with the added capability of speech.

It is interesting to speculate that the development of man's upright stance may have been responsible for an important change that took place in his throat. Among animals with a highly developed sense of smell there is a continuous airway from lungs to nose. This keeps the nasal passages continuously open, allowing the animal to detect danger even when he is eating. Man's new arrangement (a lowering of the epiglottis so that it no longer joined with the soft palate to produce a continuous airway) enabled him to close off the nasal passages completely. Thus he was able to move all his exhaled air over the tongue and lips. The change also provided an enlarged area for the amplification of vibrations from the true vocal cords. In other words, although he lost one kind of defense mechanism—a continuous sense of smell—he gained another—a more flexible form of vocal communication. The evolutionary history of animals tells us that when such a change takes place, the later mechanism is invariably the more useful one.

So we see that human communication, now the basis for man's highest achievements—in literature, science, politics—once was his major form of protection.

Words, in other words, have probably been man's greatest defense mechanism—and in more ways than one. The English philosopher Bertrand Russell has pointed out that "no matter how eloquently a dog may bark, he cannot tell you that his parents were poor but honest." Man's great capability along the lines of communication brought with it another advantage, although in this case a somewhat more dubious one. He can lie, which no animal can do. And to compound the problem, he can lie to himself as well as to others.

Let us consider the significance of this in terms of our subject, defense mechanisms. Civilized man no longer has to fear predators, at least not the animal type. He has, for the first time in history, made his house "safe from tigers." He is the only creature who can, and has, changed the environment in which he lives to any appreciable extent. And he has created a world where intelligence is more important than sheer physical strength. This is why he doesn't have to be as strong as an elephant, climb as well as a monkey, or run as fast as a cheetah.

But the life of the mind is not all peace and contentment either. Other kinds of pressure are brought to bear; if we no longer need fear being eaten, we have become subject to such other "dangers" as guilt, embarrassment, and loss of self-respect. When chickens are put together in a pen or cows in a pasture, fighting generally breaks out. Farmers expect this and are not overly concerned, for they know that once an order of dominance and submission has been established, all will be well—and that the animals are not likely to hurt each other seriously. Man appears not to have instinctive barriers against hunting his fellows. What barriers there are appear to be learned, not innate.

When things get too tough some people do what the turtle does, but in a rather modified way. They crawl into a mental, not physical, shell, and for them the outside world no longer exists.

This is a very extreme method and one that is, fortunately, not often taken. More often, a person in this kind of trouble will try to change, in his mind, the content of the world about him. A common example of this is excessive daydreaming. This is, after all, one way of answering the challenge (real or imagined) of the world in which we live. And it can appropriately be called a defense mechanism if we recall our original definition of a defense mechanism, namely, a way for a living organism to cope with the challenge presented by the world in which he lives. That man's challenges are largely mental ones and that his responses should be those of the mind reflect how far we have come since our departure from our animal forebears.

Suggested Readings

BOOKS

Adaptation, B. Wallace and A. M. Srb. Englewood Cliffs, N.J.,
Prentice-Hall, Inc., 1961 (paperback).

The Amazing Seeds, R. E. Hutchins. New York, Dodd, Mead
& Co., 1965.

Animal Clothing, G. F. Mason. New York, William Morrow
and Company, Inc., 1955.

Animal Communication, H. and M. Frings. New York, Blais-
dell Publishing Co., 1964.

Animal Conflict and Adaptation, J. L. Cloudsley-Thompson.
Chester Springs, Pa., Dufour Editions, Inc., 1965.

Animal Diversity, E. D. Hanson, Englewood Cliffs, N.J.,
Prentice-Hall, Inc., 1961 (paperback).

Animal Language, J. Huxley and L. Koch. New York, Grosset
& Dunlap, Inc., 1964 (includes record of animal sounds;
first published in 1938).

Animal Weapons, G. F. Mason. New York, William Morrow
and Company, Inc., 1949.

Animals in Armor, C. J. Hylander, New York, The Macmil-
lan Company, 1954.

Animals in Disguise, J. Poling. New York, W. W. Norton &
Co., Inc., 1966.

Animals of East Africa, C. A. Spinage. Boston, Houghton
Mifflin Co., 1963.

Audubon's Wildlife, J. J. Audubon, E. W. Teale, ed. New

York, The Viking Press, Inc., 1964.

Behavioral Aspects of Ecology, P. H. Klopfer. Englewood Cliffs, N.J., Prentice-Hall, Inc., 1962 (advanced).

Cells and Societies, J. T. Bonner. Princeton, N.J., Princeton University Press, 1955.

Curiosities of Animal Life, M. Burton. New York, Castle Books, Inc., 1966.

Insects, R. E. Hutchins, Englewood Cliffs, N.J., Prentice-Hall, Inc., 1966.

The Evolution of Life, E. E. Olsen. New York, New American Library, 1965 (paperback).

Instinct and Intelligence, S. A. Barnett. Englewood Cliffs, N.J., Prentice-Hall, Inc., 1967.

King Solomon's Ring, K. Lorenz. New York, Thomas Y. Crowell Company, 1952 (also published in paperback by Apollo).

The Mammals, A guide to the Living Species, D. Morris. New York, Harper & Row, Publishers, 1965.

The Mysterious Senses of Animals, V. B. Droscher. New York, E. P. Dutton & Co., Inc., 1965.

Never Cry Wolf, F. Mowat. Boston, Little Brown and Company, 1963 (also published in paperback by Dell Books).

On Aggression, K. Lorenz. New York, Harcourt, Brace & World, Inc., 1963 (also published in paperback by Bantam Books, Inc.).

On Growth and Form, D. W. Thompson, abridged edition with annotations by J. T. Bonner. New York, Cambridge University Press, 1961.

Patterns for Survival, L. J. and M. J. Milne, Englewood Cliffs, N.J., Prentice-Hall, Inc., 1967.

Poisonous Snakes of the World, Bureau of Medicine and Surgery, Department of the Navy, 1968. Available from Superintendent of Documents, Government Printing Office, Washington, D.C. 20402.

Of Predation and Life, P. L. Errington. Ames, Iowa, Iowa State University Press, 1967.

The Senses, W. von Buddenbrock. Ann Arbor, Mich., University of Michigan Press, 1958 (paperback).

The Territorial Imperative, R. Ardrey. New York, Atheneum Publishers, 1966.

The Treatment of Venomous Bites and Stings, H. L. Stahnke. Arizona, Arizona State University Publications, 1966, rev. ed.

Watchers, Pursuers, and Masqueraders: Animals and Their Vision, E. Raskin. New York, McGraw-Hill, Inc., 1964.

You and Your Senses, L. Schneider. New York, Harcourt, Brace & World, Inc., 1956 (paperback).

Zoology, A. M. Elliott. New York, Appleton-Century-Crofts, Inc., 3rd edition, 1963.

ARTICLES

"Beetles' Spray Discourages Predators," T. E. Eisner. *Natural History,* February 1966, p. 42.

"Behavior and Brains," J. Silvan. *Science World,* March 7, 1962, p. 6.

"Butterflies and Plants," P. R. Ehrlich and P. H. Raven. *Scientific American,* June 1967, p. 105.

"Darwin's Missing Evidence," (re: darkening of moths), H. B. D. Kettlewell. *Scientific American,* March 1959, p. 48.

"Defense Against Killers," (how gulls protect themselves), H. Kruuk. *Natural History,* April 1966, p. 48.

"Defense by Color," N. Tinbergen. *Scientific American,* October 1957, p. 48.

"Defensive Secretion of a Caterpillar (Papilio)," T. Eisner and Y. C. Meinwald. *Science,* December 24, 1965, p. 1733.

"Desert Ground Squirrels," G. A. Bartholomew and J. W. Hudson. *Scientific American,* November 1961, p. 107.

Defense Mechanisms

"Ecology of Desert Plants," F. W. Went. *Scientific American,* April 1955, p. 68.

"Electric Fishes," H. Grundfest. *Scientific American,* October 1960, p. 115.

"The Fighting Behavior of Animals," I. Eibl-Eibesfelt. *Scientific American,* December 1961, p. 112.

"The Fragile Breath of Life," several articles in *Saturday Review,* May 7, 1966.

"Heat, Cold and Clothing," J. B. Kelley. *Scientific American,* February 1956, p. 109.

"How Animals Change Color," L. J. and M. J. Milne. *Scientific American,* March 1952, p. 64.

"How to Cope With Cold Weather," A. Hamilton, *Science Digest,* December 1965, p. 56.

"The Indestructible Hydra," N. J. Berrill. *Scientific American,* December 1957, p. 118.

"Insect Assassins," J. S. Edwards. *Scientific American,* June 1960, p. 72.

"Insect's Scales are Asset in Defense," T. Eisner. *Natural History,* June/July 1965, p. 27.

"Insects Speak in Chemicals," B. Tufty. *Science News,* October 8, 1966, p. 271.

"Interaction Among Virus, Cell, and Organism," A. Lwoff, *Science,* May 27, 1966, p. 1216.

"Interferon," A. Isaacs. *Scientific American,* May 1961, p. 51.

"It Could be a 'Froad'; or Maybe a 'Trog' " (about the spadefoot), A. O. Wasserman. *Natural History,* April 1966, p. 18.

"The Leap of the Grasshopper," G. Hoyle. *Scientific American,* January 1958, p. 30.

"Mimicry," M. Rothschild. *Natural History,* February 1967, p. 44.

"Opossums: The Hardy Marsupials," *The Sciences,* October 1966, p. 21.

"Overspecialization Threatens Trees' Survival," S. Carlquist. *Natural History*, October 1965, p. 39.

"Predation's Impact on Penguins," W. J. Maher. *Natural History*, January 1966, p. 43.

"Predators and Scavengers," R. D. Estes. *Natural History*, February 1967, p. 20, and March 1967, p. 38.

"The Regeneration of Body Parts," M. Singer, *Scientific American*, October 1958, p. 79.

"Secret Weapons of Survival," (the body's defenses against diseases), R. Campbell and N. Genet. *Life Magazine*, February 18, 1966, p. 62.

"Smell and Taste," S. J. Haagen-Smit. *Scientific American*, March 1952, p. 28.

"Some Points About Cacti," *The Sciences*, May 1967, p. 13.

"The Spider and the Wasp," A. Petrunkevitch. *Scientific American*, August 1952, p. 20.

"The Unicorn in the Pool," (about the narwhal whale), N. J. Kempner. *Harper's Magazine*, November 1965, p. 57.

"Why Do Animals Fight?" J. George. *Audubon Magazine*, January/February 1966, p. 18.

Index

Index

Index

Heat-eye, 90
Hedgehog, 47
Hibernation, 97, 99, 100
Hiding, 11–28
Histamine, 62
Horns, 68, 70
  caterpillar, 70–71
  deer, 70
  rhinoceros, 68
Human body, defensive system of, 56
Hydra, 41, 42, 44
Hyenas, 113–114

Impala, 34
Insects, 125
Interferon, 61

Jumping ability, 31–32, 33, 34

Lancaster, Jane, 134
Leaping, 32–33, 34, 38
Life, basic ingredient of, 8
Lions, 118–119
Longworm, 42
Lysozyme, 56

Mammals
  archaic, 95
  egg-laying, 96
Man, comparison of, with animals, 135–138
Mice, 125–126
Microorganisms (microbes), 6
  enemy, 56–59
Mimicry, 24–25, 128
Model, 24
Mold, 54
Moles, 15
Mollusks, 50
Moths, 45
  auditory organs of, 89

clear-wing, 25
peppered, 91–92
smelling ability of, 88
Mucous membrane, 57
Mutation, 91–92, 94

Natural selection, 91–92, 94
Night creatures, 27
Nile catfish, 18

Octopus, 15–17, 20
Oysters, 50

Pain, 79–80
Paramecium, 8–10, 41
Passenger pigeon, 106, 107
Penicillin, 54, 61
"Playing possum," 51–52, 127
Poisonous substances, plant, 62
Porcupine, 45–46
Praying mantis, 31, 81
Predators, 11, 17, 23, 34, 52, 81, 87, 103, 107, 127
Prey, 3, 11, 16, 87, 103
Pronghorn, 35
Protective coloring, 18, 24, 92
  skunks, 2
  See also Mimicry
Protozoa, 6, 43
Puffer, 48
Puma, 34

Quills, 45, 46

Rabbit, 3
Rainfall, 53–54
Regenerative powers, 41–44
Reproduction, 8
  as a defense mechanism, 10
  successive, 42
Reptiles, 95–97, 103
Resistance, environmental, 10

149

Index

ABOUT THE AUTHOR

Hal Hellman's wide range of interests is reflected in his education and experience. He holds a master's degree in physics, as well as degrees in economics and industrial management. He has published numerous articles (including booklets for the Atomic Energy Commission on lasers and spectroscopy) and nine books for young adults on such varying scientific subjects as color and high-energy physics. Mr. Hellman, his wife, and two daughters presently live in Leonia, New Jersey.